The Greate Invention of Algebra

Doe you not here startle, to see every day some of your inventions taken from you; for I remember long since you told me as much (as Kepler has just published) that the motions of the planets were not perfect circles. So you taught me the curious way to observe weight in Water, and within a while after Ghetaldi comes out with it in print. A little before Vieta prevented you of the Gharland for the greate Invention of Algebra. Al these were your deues and manie others that I could mention; and yet too great reservednesse hath rob'd you of these glories ... Onlie let this remember you, that it is possible by to much procrastination to be prevented in the honor of some of your rarest inventions and speculations.

William Lower to Thomas Harriot, 6 February 1610

The Greate Invention of Algebra

THOMAS HARRIOT'S TREATISE ON EQUATIONS

Jacqueline A. Stedall

Clifford Norton Student in the History of Science,
The Queen's College, Oxford,
and
Open University

OXFORD
UNIVERSITY PRESS

UNIVERSITY PRESS

Great Clarendon Street, Oxford OX2 6DP

Oxford University Press is a department of the University of Oxford.
It furthers the University's objective of excellence in research, scholarship,
and education by publishing worldwide in

Oxford New York

Auckland Bangkok Buenos Aires Cape Town Chennai
Dar es Salaam Delhi Hong Kong Istanbul Karachi Kolkata
Kuala Lumpur Madrid Melbourne Mexico City Mumbai Nairobi
São Paulo Shanghai Taipei Tokyo Toronto

Oxford is a registered trade mark of Oxford University Press
in the UK and in certain other countries

Published in the United States
by Oxford University Press Inc., New York

A catalogue record for this title is available from the British Library

Library of Congress Cataloging in Publication Data

Stedall, Jacqueline.
The greate invention of algebra: Thomas Harriot's treatise on equations/Jacqueline A. Stedall.

Includes bibliographical references and index.

1. Equations   2. Algebra–England–History–17th century   3. Harriot, Thomas,
1560-1621–Influence   I. Harriot, Thomas, 1560-1621   II. Title.

QA211 .S77 2003   512.9'4–dc21   2002042558

ISBN 0 19 852602 4 (acid-free paper)

10 9 8 7 6 5 4 3 2 1

Typeset by Cepha Imaging Pvt. Ltd.
Printed in Great Britain
on acid-free paper by Biddles Ltd, Guildford & King's Lynn

Dedication

For Robert Fox, John North
and Peter Neumann

Acknowledgements

❧

I am indebted to a number of librarians and archivists who have helped me to locate and use the manuscripts of which this edition is based, first and foremost to Colin Harris and his staff at the Bodleian Library, Oxford, who have so often and so readily provided me with the hefty boxes containing the photocopies of the Harriot and Cavendish papers. I am also grateful to Alison McCann of the West Sussex Record Office for her assistance with the Petworth papers and to Lord Egremont for his permission to copy and use them. Sarah Wickham of Lambeth Palace Library helped with great efficiency with the Torporley manuscripts, and the archivists of the Northamptonshire Record Office with the Isham papers.

I would like to give special thanks also to Muriel Seltman who first introduced me to Harriot's manuscripts, and who has given me much warm support, and to Stephen Clucas who helped me in researching details of Harriot's life and acquaintances.

While preparing this edition I have held the Clifford Norton Studentship in the History of Science at The Queen's College, Oxford, and have been a member of the Centre for the History of the Mathematical Sciences at the Open University. I have dedicated this book to three people who have taken a particular interest in this project and who have given me unstinting encouragement and support.

The Queen's College, Oxford J.A.S.
April 2002

Contents

APPENDIX

Illustrations

Figures 1, 2, 4, 5, 7, 8, 10, 11, 12, 13 are published by permission of the British Library; Figure 6 by permission of Lambeth Palace Library; Figure 9 by permission of Lord Egremont.

Introduction

Introduction

T homas Harriot (c.1560–1621) was an innovative thinker and practitioner in several branches of the mathematical sciences: navigation, astronomy, optics, geometry and algebra, but he never published any of his scientific or mathematical findings.[1] At his death he left behind over four thousand manuscript sheets of observations, calculations and diagrams, but the range of the contents and the disorder of the papers have defeated almost all subsequent attempts at publication.[2] For this edition I have ordered, translated and annotated about one hundred and forty of the sheets, those concerned with Harriot's investigations into the solution and structure of polynomial equations. In the years immediately after Harriot's death some of this material was published by Harriot's colleagues and Executors in the *Artis analyticae praxis* (1631), but the editors selected and reordered Harriot's work in such a way that the *Praxis* often bears little resemblance to the manuscripts, and fails to do full justice to the quality and originality of Harriot's insights. The relevant sheets are now separated and dispersed amongst the surviving manuscripts, but Harriot's pagination, the mathematical content and other contemporary evidence allow them to be reassembled in what appears to be the original sequence. The result is a lucid and self-contained *Treatise on equations*, published here for the first time in its original form.

I. The *Treatise on equations*

Harriot, Torporley and Viète

Nothing is known of Thomas Harriot's early background. The appearance of his name in the Oxford University Register in 1577 suggests that he was born about 1560, and the entry implies that he already lived in Oxford, and that he took up residence in St Mary's Hall, affiliated to Oriel College. Some time after Harriot's graduation around 1580 he entered the service of Walter Ralegh, and was employed by him as navigator and scientist on a voyage to north America from 1585 to 1586.[3] Harriot's report of the expedition, *A briefe and true report of the new found land of Virginia* (1588), was the only book he ever published.

Harriot's reputation as a mathematician became established in the years after his return from America. In 1593 Gabriel Harvey called him a 'profounde Mathematician' (alongside Thomas Digges and John Dee), and in 1594 Robert Hues announced in his *Tractatus de globis* that a further treatise could be expected from the 'mathematician and philosopher, Thomas Harriot'.[4] From the early 1590s Harriot had a lifelong patron and benefactor in Sir Henry Percy, the ninth earl of Northumberland. The Earl was imprisoned in the Tower of London from 1605 to 1621 on suspicion of involvement in the Gunpowder Plot (his cousin Thomas Percy was one of the ringleaders) but he continued to maintain Harriot at his London home, Syon House at Isleworth in Middlesex. Most of Harriot's mathematical and scientific companions were also in some way connected with the Earl's household:[5] Walter Warner (*c.* 1557–1643) was keeper of Percy's library and scientific instruments, while Robert Hues (1553–1632), a friend of Harriot's since his time at Oxford, became tutor from 1615 to the Earl's sons. Harriot discussed algebra, optics and astronomy with William Lower (1570–1615), keeper of Percy's Welsh Estates, and he made astronomical observations with Lower's successor John Protheroe (1582–1624). Through Protheroe Harriot also became acquainted with Thomas Aylesbury (1580–1657).

Another intellectual companion was Nathaniel Torporley (1564–1632), who had graduated from Oxford four years after Harriot. It is not known when Harriot and Torporley first met but both were interested in mathematics. Torporley wrote to Harriot from Paris on the eve of his first meeting with Viète, the self-styled 'French Apollonius':[6]

> I am gathering up my ruined wittes, the better to encounter that French Apollon: if it fortune that either his courtsie or my boldnes effecte our conference; tomorrow beinge the daye, when I am appoynted by his Printer, as little Zacheus to climbe the tree, to gayne a view of that renoumned analist. What after followes in [his] presence I hope shortly to relate . . .

What followed was that Torporley became Viète's amanuensis,[7] and it was almost certainly through Torporley that Harriot acquired his detailed know-ledge of Viète's mathematics: there is, for example, among Harriot's papers a sheet headed: 'A proposition of Vietas delivered by Mr. Thorperly',[8] and there are numerous other references to Viète's mathematics throughout the manuscripts. In particular, Viète's algebra became the foundation of Harriot's algebra, and so we need to consider briefly what Viète himself contributed to the development of the subject.

From the Arabic expositions of the ninth century to the European text-books of the late sixteenth century, algebra was understood to be the practice of forming and solving polynomial equations: quadratics, cubics and, occasionally, quartics. Coefficients in those equations were always regarded as positive, leading to several apparently different cases which had to be handled separately. Solutions, in the form of a list of instructions, were given for specific equations, it being understood that any other equation of the same type could be solved in a similar way. Geometrical arguments were sometimes appended to justify the methods, but algebra was seen essentially as advanced arithmetic, an arithmetic in which numbers were replaced by symbols. The earliest notation, R, Z, C, for root, square, cube, was hardly more than a system of abbreviation that gave little or no insight into mathematical structure.

Viète transformed both the form and the content of traditional algebra. He used the vowels A, E, ... to represent unknown quantities, with A *quadratus*, A *cubus*, etc. for A-squared, A-cubed, and so on, so that both the base quantity and its power were immediately clear. Known or given quantities were denoted by consonants B, C, D, ... though to maintain dimensional homogeneity Viète introduced terms like B *plane*, Z *solid*, so his writing often appears more verbal than symbolic. Nevertheless, in Viète's notation equations could be written for the first time entirely in letters, or 'species', that is in completely general form. Viète was concerned not only with solutions, but with the underlying structure of equations and, immersed in classical Greek mathematical thought, he saw that structure in terms of proportion:[9]

> If B *plane* times A minus A *cubed* is equal to Z *solid*, then B *plane* is composed of the squares of three proportionals; and Z *solid* is made by the multiplication of one of the extremes by the sum of the squares of the other two, making A the first or the third [of the proportionals].
>
> Suppose the proportionals are 2, $\sqrt{20}$, 10, then it may be said that 124N–1C is equal to 240, and 1N is 2 or 10.

Thus classical Greek mathematics with its emphasis on the use of proportion helped to create Viète's perception of algebraic structure. But the new availability of Greek texts influenced European mathematics in other ways too, and in particular began to extend the boundaries of algebraic investigation.[10] Both Rafael Bombelli and Viète yoked algebra to the arithmetic of Diophantus,[11] but Viète went further and recognized as no one had before him, that algebra could also be applied to geometrical problems. He saw that his 'species', or symbols, could represent not just numbers but geometrical quantities: lines, planes and solids, so that geometry, previously called upon to justify algebra, now became itself amenable to algebraic treatment. Indeed, Viète supposed that the algebraic method was the analytic tool by which classical geometrical theorems had originally been discovered, and so for him, and for those who followed him, algebra became the 'analytic art', the means by which known results could be verified and new theorems found.[12] Viète's introduction to algebra, *In artem analyticem isagoge* of 1591, introduced the fundamental definitions and rules of the subject, hinted at the scope of the methods, and ended with Viète's vision of a new mathematical future: 'the analytic art', he wrote, 'claims for itself the greatest problem of all, *to leave no problem unsolved.*'[13]

To solve every problem in mathematics meant carrying every algebraic analysis to its conclusion, which in turn meant solving whatever equations might arise. Viète dealt with techniques of solving equations in two separate treatises, one theoretical and one practical. The theoretical text, *De aequationem recognitione et emendatione tractatus duo*, dealt with the standard techniques of handling equations up to the fourth degree. Though not published until 1615, it was written in the early 1590s,[14] and some of the material in it might well have been known to Harriot through Torporley before it appeared in print. The practical treatise, *De numerosa potestatum ad exegesin resolutione*, dealt with solving equations by numerical methods, the first exposition in Europe of the solution algorithms developed by twelfth-century Islamic mathematicians.[15] *De potestatum resolutione* was published in 1600, but as with *De aequationum recognitione*, the principles were known to Viète much earlier, certainly by 1591.[16] Harriot studied the text of *De potestatum resolutione* in detail and, as he had done with the *Isagoge*, re-wrote much of the material in his own notation. More importantly, he explored for himself the theoretical underpinning of Viète's method, and so developed his own treatment of the structure and solution of polynomial equations. The re-written material from Viète, and Harriot's theoretical treatment, together form a self-contained, but untitled, treatise. I have called it the *Treatise on equations*, and in this edition it is published complete for the first time.

The *Treatise on equations* is undated but much of it arises so directly from *De potestatum resolutione* that it was probably written shortly after that book appeared in 1600.[17] Some of it (parts of Sections (*d*), (*e*) and (*f*), perhaps) may have been written earlier, and may have been what William Lower had in mind when in 1610 he lamented Harriot's failure to publish his 'greate Invention of Algebra':

> So you taught me the curious way to observe weight in Water, and within a while after Ghetaldi comes out with it in print. A little before Vieta prevented you of the Gharland for the greate Invention of Algebra. Al these were your deues and manie others that I could mention; and yet too great reservednesse hath rob'd you of these glories ...[18]

Harriot's notation

Harriot's notation, very different from Viète's, was essential to his new understanding of equations. His most important innovation was undoubtedly his use of *ab* to represent *a* multiplied by *b*, and consequently *aa*, *aaa* for what is now written a^2, a^3, etc. He was not the first to use such a system: as early as 1544 Michael Stifel had written 27AB for the product of 3A and 9B,[19] but introduced A, B, C, etc. only when a problem dealt with more than one unknown quantity. Almost all sixteenth-century algebra textbooks restricted themselves to linear and quadratic equations in a single unknown, which could be handled with a few special signs for roots and squares, so Stifel's notation was hardly ever necessary and was not generally adopted.

When first Bombelli and then Simon Stevin took up Cardano's treatment of cubic and quartic equations they adopted a system in which powers were represented by encircled numbers so that, for example, '6③ plus 3②' represented what would now be written as $6x^3 + 3x^2$, a notation that is unambiguous for a single unknown.[20] Viète took the next important step, in the *Isagoge* in 1591, by introducing the letters A, E, etc. for unknown quantities but ignored the advances made by Bombelli and Stevin, and described powers verbally as *A quadratus*, *A cubus*, and so on. Harriot retained Viète's use of vowels for unknown quantities but changed to lower case *a*, *e*, etc. and represented powers by repeated multiplication. This is occasionally tedious to write but never to read; indeed, it sometimes displays mathematical structure more clearly than does modern superscript notation, and was ideally suited to Harriot's purpose of exploring polynomial equations in terms of their coefficients.

To denote equality Harriot adopted a variation of the $=$ sign devised by Robert Recorde in *The whetstone of witte* in 1557.[21] In Harriot's manuscripts the symbol always appears with two short cross strokes between the horizontals, but these disappeared when some of his algebra was later put into print. Harriot also introduced two new signs: $<$ and $>$. In manuscript these are longer and more curved than the modern printed versions, and are written with two cross strokes over the open end (see fig. 1).[22] Harriot used all three symbols, $=$, $<$, $>$ both horizontally and vertically as occasion demanded, working across or down the page at will.

Harriot indicated parenthesis by a single comma, as in: $\sqrt{,}\ bb - cc$ or $b -, c - d = b - c + d$. He also, however, often used commas to denote the multiplication of a letter by a number, as in $2,a$. Multiplication of compound quantities was shown by enclosing them in a right-angled bracket:

$$\left.\begin{array}{r} a+b \\ b+c \end{array}\right| = \begin{array}{l} ab + bb \\ + ac + bc \end{array}$$

with the steps of the multiplication carried out exactly as in a long multiplication in arithmetic. Division was represented in the usual way by writing the dividend and divisor above and below a horizontal line, as in fractions.[23] Harriot always used the by then well established (but by no means universal) symbols $+$ and $-$ to denote addition and subtraction,[24] but also introduced two new symbols: $\pm$ and $\mp$ to denote alternative possibilities.[25] This enabled him to handle two, or even four, equations at a time, a useful saving when every combination of signs had to be dealt with separately.[26]

The *Operations of arithmetic in letters*

Harriot wrote a four-page introduction to his notation, entitled *Operations of arithmetic in letters*,[27] in which he demonstrated the use of lower case letters in examples of addition, subtraction, multiplication and division, and in the standard rules for simplifying equations. The examples on the first two sheets were Harriot's own, but those on the third sheet, on division (see fig. 2), were taken directly from the *Isagoge*.[28] A comparison of corresponding statements from Viète and Harriot demonstrates immediately the conciseness and clarity of Harriot's notation:

Viète:[29]   If to $\dfrac{A\ plane}{B}$ there should be added $\dfrac{Z\ squared}{G}$,

the sum will be $\dfrac{G \text{ times } A\ plane + B \text{ times } Z\ squared}{B \text{ times } G}$

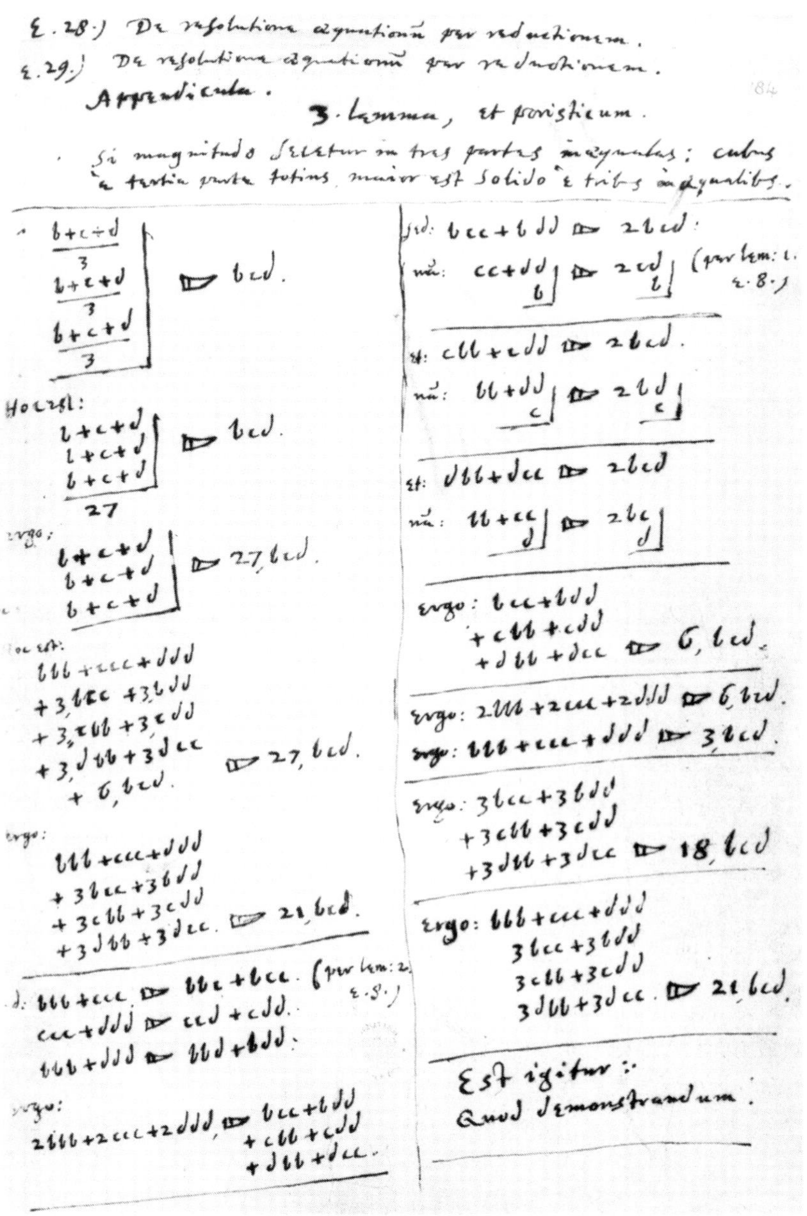

Fig. 1 Sheet *e*.29) of the *Treatise on equations*, Add MS 6783, f. 184.

3·)

$$\frac{ba}{b} = a.$$

$$\frac{bca}{b} = ca.$$

$$\frac{bca}{c} = ba.$$

$$\frac{ba}{c} + z = \frac{ba+zc}{c} = \frac{ba+zc}{c}$$

$$\frac{a\varepsilon}{b} + z = \frac{a\varepsilon+zb}{b}$$

$$\frac{ac}{b} + \frac{zz}{g} = \frac{acg+bzz}{bg} \quad\frac{}{bg} = \frac{acg+bzz}{bg}$$

$$\frac{ac}{b} - z = \frac{ac-zb}{b} = \frac{ac-zb}{b}$$

$$\frac{ac}{b} - \frac{zz}{g} = \frac{acg}{bg} - \frac{zzb}{bg} = \frac{acg-zzb}{bg}$$

$$\left.\frac{ac}{b}\right\}_{b} = \frac{acb}{b} = ac. \qquad \left.\begin{array}{c}\frac{ac}{b}\\[4pt]\frac{zz}{g}\end{array}\right\} = \frac{aczz}{bg}$$

$$\left.\begin{array}{c}\frac{ac}{b}\\[4pt]z\end{array}\right| = \frac{acz}{b}$$

$$\left.\begin{array}{c}\frac{aaa}{b}\\[4pt]d\end{array}\right| = \frac{aaa}{bd} \quad . \; vel: \; \frac{\frac{aaa}{b}}{d}\left| = \frac{\frac{aaa}{b}}{\frac{db}{b}}\right| = \frac{aaa}{db}$$

$$\left.\begin{array}{c}bg\\[4pt]\frac{ac}{d}\end{array}\right| = \frac{bgd}{ac} \quad . \; vel: \; \frac{bg}{\frac{ac}{d}}\left| = \frac{\frac{bgd}{d}}{\frac{ac}{d}}\right| = \frac{bgd}{ac}$$

$$\left.\begin{array}{c}\frac{bbb}{z}\\[4pt]\frac{aaa}{dc}\end{array}\right| = \frac{bbbdc}{zaaa} \; . \; vel \; \frac{\frac{bbb}{z}}{\frac{aaa}{dc}}\left| = \frac{\frac{bbbdc}{zdc}}{\frac{aaaz}{dcz}}\right| = \frac{bbbdc}{zaaa}$$

Fig. 2 Sheet 3) of *Operations of arithmetic in letters*, Add MS 6784, f. 325.

Harriot:[30] $\dfrac{ac}{b} + \dfrac{zz}{g} = \dfrac{acg + bzz}{bg}$

(Note that Harriot adhered to Viète's dimensional homogeneity, by replacing *A plane* by the two-dimensional quantity *ac*.) Another example:

Viète:[31] If *B* times *G* should be divided by $\dfrac{A\ plane}{D}$,

both magnitudes having been multiplied by *D*,

the result will be $\dfrac{B\ \text{times}\ G\ \text{times}\ D}{A\ plane}$

Harriot:[32] $\dfrac{bg}{\dfrac{ac}{d}} = \dfrac{bgd}{ac}$

On the fourth sheet of *Operations of arithmetic in letters,* Harriot turned to the standard rules for simplifying equations: moving terms from one side to another, reducing the leading coefficient to one, and dividing out excess powers of the unknown.[33] Viète gave these three rules the names of *antithesis, hypobibasmus* and *parabolismus* respectively,[34] and Harriot retained both the names and Viète's examples (see figs. 3 and 4):

Viète:[35] *A squared* minus *D plane* is supposed equal to *G squared* minus *B* times *A.* I say that *A squared* plus *B* times *A* is equal to *G squared* plus *D plane* and that by this transposition and under opposite signs of conjunction the equation is not changed.

Harriot:[36] Suppose $aa - dc = gg - ba$
I say that $aa + ba = gg + dc$ by *antithesis.*

Once again, the lucidity and economy of Harriot's notation is obvious. Easy to read and easy to use, it reveals algebraic structure and acts as an aid to thinking in a way that Viète's verbal descriptions can not.

The *Treatise on equations*

Harriot's *Treatise on equations* was written in six sections. The last five of these were lettered by Harriot as (*b*) to (*f*) and so I have taken the natural step of giving the first the letter (*a*). Sections (*a*) to (*c*) deal with the numerical solution of equations and are closely based on Viète's *De potestatum resolutione.* Section (*d*) demonstrates the multiplicative structure of polynomials.[37]

9 Atque idcirco ſi accidat homogeneum ſub data menſura immiſceri homogeneo ſub gradu, fiat Antitheſis.

Antitheſis eſt cum adficientes affectæve magnitudines ex una æquationis parte in alteram tranſeunt ſub contraria adfectionis nota. Quo opere
æqualitas non immutatur. Id autem obiter eſt demonſtrandum.

P r o p o s i t i o I.
Antitheſi æqualitatem non immutari.

Proponantur A quadratum minus D plano æquari G quadrato minus B in A. Dico A
quadratum plus B in A æquari G quadrato plus D plano, neque per iſtam tranſpoſitionem
ſub contraria adfectionis nota æqualitatem immutari. Quoniam enim A quadratum minus D plano æquatur G quadrato minus B in A addatur utrobique D planum plus B in
A. Ergo ex communi notione A quadratum, minus D plano plus D plano plus B in A æ-
quatur G quadrato, minus B in A, plus D plano: plus B in A. Iam adfectio negata in eadem
æquationis parte elidat adfirmatam: illic evaneſcet adfectio D plani, hic adfectio B in A,
& ſupererit A quadratum plus B in A æquale G quadrato plus D plano.

10 Et ſi accidat omnes datas magnitudines duci in gradum, & idcirco homogeneum ſub data omnino menſura non ſtatim offerri , fiat Hypobibaſmus.

Hypobibaſmus eſt æqua depreſſio poteſtatis & parodicorum graduum
obſervato ſcalæ ordine , donec homogeneum ſub depreſſiore gradu cadat
in datum omnino homogeneum cui comparantur reliqua. Quo opere æ-
qualitas non immutatur. Id autem obiter eſt demonſtrandum.

**Hypobibaſmi opus à Paraboliſmo differt in eo tantum quod per Hypobibaſmum utraque æqualitatis
pars ad quantitatem ignotam adplicatur; per Paraboliſmum vero ad quantitatem certam , ut ex exem-
plis ab authore allatis perſpicuum eſt.**

P r o p o s i t i o II.
Hypobibaſmo æqualitatem non immutari.

Proponatur A cubus, plus B in A quadratum; æquari Z plano in A. Dico per hypobibaſmum A quadratum, plus B in A; æquari Z plano.

Illud enim eſt omnia ſolida diviſiſſe per communem diviſorem , à quo non immutari
æqualitatem determinatum eſt.

11 Et ſi accidat gradum altiorem , ad quem adſcendet quæſita magnitudo,
non ex ſe ſubſiſtere, ſed in aliquam datam magnitudinem duci, fiat Paraboliſmus.

Paraboliſmus eſt homogeneorum , quibus conſtat æquatio , ad datam
magnitudinem, quæ in altiorem quæſititiæ gradum ducitur, communis ad-
plicatio ; ut is gradus poteſtatis nomen ſibi vendicet, & ex ea tandem æquatio ſubſiſtat. Quo opere æqualitas non immutatnr. Id autem obiter eſt
demonſtrandum.

P r o p o s i t i o III.
Paraboliſmo æqualitatem non immutari.

Proponatur B in A quadratum plus D plano in A æquari Z ſolido. Dico per Parabo-
liſmum A quadratum plus $\frac{D \text{ plano}}{B}$ in A æquari $\frac{Z \text{ ſolido}}{B}$ Illud enim eſt omnia ſolida diviſiſſe per B communem diviſorem , à quo non immutari æqualitatem determinatum eſt.

12 Et tunc diſerte exprimi æqualitas cenſetor & dicitor ordinata : ad Ana-
logiſmum, ſi placet , revocanda , tali præſertim cautione ; ut ſub extremis
facta, tum poteſtati tum adfectionum homogeneis reſpondeant; ſub mediis
vero, homogeneo ſub data menſura.

13 Vnde etiam Analogiſmus ordinatus definiatur ſeries trium quatuorve

B

magni-

Fig. 3 Page 9 of Viète's *Isagoge* (1646 edition).

Fig. 4   Sheet 4) of *Operations of arithmetic in letters* Add MS 6784, f. 324.

Sections (*e*) and (*f*) are systematic treatments of cubic and quartic equations respectively. In more detail the contents are as follows:

Section (a): On solving equations in numbers

Harriot's Section (*a*) is based on Problems 1 to 6 of Viète's *De potestatum resolutione*, and treats equations that are positively affected, that is, of the form $a^n + ba^m = c$ (with $b > 0$, $c > 0$ and $n > m$). Such equations have just one positive root, and Viète set out in each case the algorithm for finding it. In his Section (*a*) Harriot re-wrote Viète's Problems 1 to 6 in his own notation, and added some further examples of his own. (See the Appendix for a detailed comparison between Harriot's material and Viète's.)

Section (b): On solving equations in numbers

Harriot's Section (*b*) takes up Problems 10 to 15 of Viète's *De potestatum resolutione*, and is concerned with equations negatively affected, that is, of the form $a^n - ba^m = c$ (with $b > 0$, $c > 0$ and $n > m$). Such equations, like those in Section (*a*) have just one positive root. Again, Harriot set out Viète's algorithm for each problem in his own notation.

Section (c): On solving equations in numbers

Section (*c*) covers Problems 16 to 20 of Viète's *De potestatum resolutione*. Here the equations are of the form $ba^m - a^n = c$ (with $b > 0$, $c > 0$ and $n > m$), which always have either two positive roots or none. Viète's equations were chosen to have two positive roots and he solved for both, and also explained verbally how one root might be found from the other. Harriot again wrote out Viète's solutions and explanations in his own notation, but now began also to carry out investigations of his own into the relationships between roots and coefficients. For equations of each degree (up to and including quartics) he was able to determine bounds within which the (positive) roots must lie, and expressed those bounds by means of double inequalities using his new symbols, < and >.

Section (d): On the generation of canonical equations

In Section (*c*) Harriot had given examples of canonical equations, or representative forms, in which it was clear how the coefficients were related to the roots. For example, the canonical equation $bc = ba + ca - aa$ has two positive roots, $a = b$ and $a = c$. In Section (*d*) Harriot showed how such canonical forms arose, and in doing so made a contribution of inestimable importance to the understanding of polynomial equations.

Where Viète had analysed equations in terms of ratios, Harriot saw instead the possibility of writing polynomials as products of factors of lower degree.[38] He began with the multiplication $(a - b)(a - c)$ (see fig. 5) where a as usual is the unknown quantity. Then by systematically changing signs and adding further factors he built up a series of quadratic, cubic and quartic polynomials, and noted the equations that arose from setting those polynomials to zero. He used factors only of the form $(a \pm b)$, $(aa \pm df)$ or, occasionally, $(aaa \pm dfg)$,

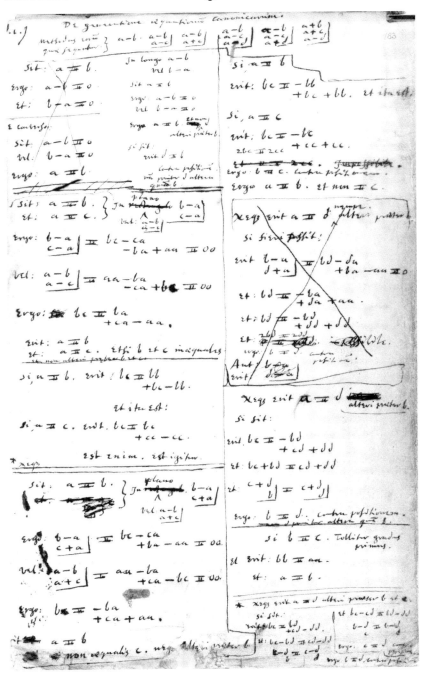

Fig. 5   Sheet $d.1$) of the *Treatise on equations*, Add MS 6783, f. 183.

and in the resulting equations, the relationships between the roots and the coefficients are immediately clear, especially in Harriot's layout where terms containing the same power are grouped vertically.

Harriot realized that certain conditions between the roots would cause one or more powers to vanish from the equation: in the equation $aa - ba + ca - bc = 0$, for instance, the linear terms disappear if $b = c$, and the equation reduces to $aa - bb = 0$ (which, like the original, has a positive root, $a = b$). For two powers to vanish, two independent conditions must be satisfied (one for each power). In sheets $d.10$) to $d.12$) Harriot noted the pairs of conditions required for the elimination of two distinct powers from a quartic. Solving such pairs simultaneously led him to square roots of negative quantities, and he handled these with ease and without comment.

Harriot defined pairs of conjugate equations as those in which the odd powers a, aaa, etc. have opposite signs. Thus the positive roots of any equation are the negative roots of its conjugate and it is clear from the many examples in this section that Harriot recognized this relationship.[39]

The most often repeated criticism of Harriot's work has been that he restricted himself to positive real roots.[40] Two things need to be said on this point: first, that Harriot's work on equations arose directly from his study of Viète's algorithms, which were specifically designed to elicit positive roots, and so it was natural that he should take the search for positive roots as his starting point; second, it is clear that as his work progressed, Harriot began to recognize the importance of both negative and complex roots. There are many examples of the former, and a few of the latter, in Sections (e) and (f) but Harriot also began to see how and where they should occur in Section (d). There are two additional sheets, marked $d.7.2$) and $d.13.2$), which are later additions to Section (d), and both sheets allow for both negative and complex roots.

Section (e): On solving equations by reduction
Harriot began Section (e) by systematically listing all possible cases of linear, quadratic, cubic and quartic equations, according to the signs of the coefficients.

The next part of Section (e) is Harriot's treatment of cubic equations lacking a square term. He treated three distinct forms:

$$aaa + 3bba = 2ccc \quad \text{(1 positive root)} \qquad \text{sheet } e.5)$$
$$aaa - 3bba = 2ccc \quad \text{(1 positive root)} \qquad \text{sheets } e.6) \text{ to } e.11)$$
$$3bba - aaa = 2ccc \quad \text{(0 or 2 positive roots)} \quad \text{sheets } e.12) \text{ to } e.14)$$

Harriot dealt at length with the second form, $aaa - 3baa = 2ccc$, and distinguished the three cases $c > b$, $c = b$, $c < b$, which he labelled hyperbolic, parabolic and elliptic respectively.[41] Hyperbolic equations have one positive root and a pair of complex roots; elliptic equations have one positive and two negative roots. The elliptic form gives rise to the 'irreducible case' in which the solution formula includes square roots of negative numbers, but the solution is real because the imaginary terms cancel each other out. Harriot showed that in principle such an equation, of the form $aaa - 3bba = 2ccc$, could be solved from its conjugate equation, $3bba - aaa = 2ccc$, which has the same numerical roots with different signs.

The final part of Section (e) shows how to remove the square term from any cubic equation; the equation thus reduced can then be solved by the methods outlined previously.

Section (f): On solving equations by reduction
This long section is now the most disordered of Harriot's treatise, and falls into six parts, which may have been written out at different times. The unifying factor is that all of them deal with quartic equations. One part (now labelled $f \delta$) shows how to solve a quartic equation lacking the cube term, by the method devised by Ludovico Ferrari and published by Girolamo Cardano in 1545, later systematized by Bombelli, Stevin and Viète.[42] The remaining parts give detailed instructions on how to remove the cube term from any quartic by a simple linear transformation, with numerical examples for each case. This was new: the removal of the cube term from a quartic by this method is not found in the published work of any of Harriot's predecessors. If part $f \delta$ is placed first, Section (f) follows the same overall pattern as Section (e): first, the solution of equations lacking the second highest term, followed by instructions on how to reduce any equation to such a form.

II. Harriot's algebra after 1621

A few days before his death in 1621 Harriot dictated a Will, in which he gave special attention to his mathematical papers. The relevant instructions were as follows:[43]

> ... I Thomas Harriots of Syon in the County of Midd Gentleman being
> troubled in my body with infirmities. But of perfect mind and memory
> Laude and prayse be given to Almightie God for the same do make
> and ordayne this my last will and testament...

I ordaine and Constitute the aforesaid NATHANIEL THORPERLEY
first to be Overseer of my Mathematical Writings to be received of my
Executors to peruse and order and to separate the chief of them from
my waste papers, to the end that after he doth understand them he
may make use in penning such doctrine that belongs unto them for
public uses as it shall be thought Convenient by my Executors and
him selfe And if it happen that some manner of Notations or writings
of the said papers shall not be understood by him then my desire is
that it will please him to Conferre with Mr Warner or Mr Hughes
Attendants on the aforesaid Earle Concerning the aforesaid doubt.
And if he be not resolved by either of them That then he Confer with
the aforesaid JOHN PROTHEROE Esquior or the aforesaid THOMAS
ALESBURY Esquior. (I hoping that some or other of the aforesaid four
last nominated can resolve him) And when he hath had the use of
the said papers so long as my Executors and he have agreed for the
use afore said That then he deliver them again unto my Executors to
be put into a Convenient Trunk with a lock and key and to be placed
in my Lord of Northumberlands Library and the key thereof to be
delivered into his Lordships hands.

It is clear that Harriot thought Torporley the person best fitted to understand,
transcribe and edit his mathematical papers. From 1608 Torporley was the
rector of Salwarpe in Worcestershire and so may not have been acquainted
with Harriot's later work: in this case Harriot suggested that Warner and
Hues would be in a position to assist, with Protheroe and Aylesbury (also
Executors of the Will) as the final arbiters. Preparations for carrying out
Harriot's wishes were put in train: in 1622 Torporley resigned his position
and probably moved to one of the Earl of Northumberland's residences, Syon
House in Middlesex, or Petworth House in Sussex. Protheroe paid him a
pension, and instructed his wife to continue the payments after his death
(in 1624).[44] Later, Torporley was probably supported by Henry Percy: the
Earl's household papers show a payment to Torporley in 1626, and he was at
Petworth in 1627.

The mathematical papers handed over to Torporley were carefully listed
by Aylesbury, and the list was endorsed by both Protheroe and Torporley.[45]
It was headed:

> Copyed from Mr Protheroe
> A note of the papers
> and bookes in Mr Harriot's

trunke delivered to

Mr Torporley

There were sixty items (plus nineteen more added later) and the first nine alone give some idea of the overwhelming task faced by the Executors:

Analytiques in 16 bundells

De Centro gravitatis 3 b.   b. bundells
De Jovialibus planetis

The spots in the sun
The faces of the moon   all in one great b.

Of the observations of the moon, 1 great b more

Eratosthenes Batavus de quadrilatero in circulo, de parabola

Silo princeps fecit, diluvium Noachi, generatio maris et feminae with some other papers of genealogies

3 b. On Vietaes zetetiques, with a few miscellaneous papers de Inclinationibus & porismatis (All these bound up in a pack thred together)

Of the errors in observations by Instruments which cannot be made exactly ad minutum, 1 b.

Certaine observations in a great b. most cleane paper

Torporley was faced with the same daunting quantity of material as every potential editor since, but among his surviving manuscripts is a work now known as the *Congestor* ('Compilation') in which he began the task of presenting some of Harriot's work.[46] After a lengthy preamble he identified the topics that he proposed to deal with:[47]

Numeris Compositis vel aliis primis dignoscendia	} *proponimus*
Facultate Congestita invento Hariotaeo nostroque	} *nobis ad*
De Divisione ingemminata	} *agendur*
Radicalium dispensatione Hariotae	}
Speciosa Logistica ab eodem perfecta a FV	}

identification of a number as composite or prime	} we propose
a compilation of ideas also discovered by our Harriot	} by us to be
Of repeated division	} treated
Harriot's treatment of surds	}
arithmetic in species perfected by him from Viète	}

The *Congestor* comprises the first two sections of this scheme: first, several pages on the identification of prime numbers, and related problems, ending with a nine-page table of prime factors for numbers up to 20399; second, a section headed 'Thomas Hariotus, examinatur Stifelius de numeris diagonalibus', a copy of Harriot's work on Pythagorean triples.[48] A fair copy of the *Congestor* was dedicated to the Earl of Northumberland at Petworth in October 1627.

Torporley did not complete the remainder of his plan, and it would seem that at some stage Harriot's Executors relieved him of his duties. We do not know why this happened, but Torporley later complained that 'my enemies accuse me to the Master of Petworth as being, among other things, an ignorant logic-chopper',[49] and the Executors may have been simply concerned about his age and his intellectual ability to complete the task. For whatever reason, Harriot's work on equations was eventually edited not by Torporley, as Harriot had intended, but by Walter Warner, possibly assisted by others: Torporley later remarked that one of the editors had been 'lifted to heaven',[50] and this would have been Protheroe, who died in 1624, so perhaps the Executors divided Harriot's papers even before that date.

Warner's edition was published in 1631 as the *Artis analyticae praxis*, commonly known as the *Praxis*.[51] Warner was not previously known as a mathematician, and never understood Harriot's work as well as Torporley did. Instead of editing the manuscripts as they stood, he chose to select and reorder the material in a manner that will be explained in detail below. In doing so he not only destroyed the coherence of Harriot's treatise but made it appear considerably less sophisticated than in fact it was.

The *Praxis*

The *Praxis* opens with a four-page introduction and six pages of preliminary definitions. The main text is then in two parts: Part I, subdivided into six sections, deals with the generation and reduction of equations; Part II is concerned with numerical solution.

Part I, Section 1, is virtually identical to Harriot's *Operations of arithmetic in letters*. The remainder of Part I then covers Harriot's Sections (*d*), (*e*) and (*f*), but in a way that diverges radically from the manuscripts. Warner rearranged Harriot's material, and seems to have done so to make it fit to three of the definitions that appear in the opening pages of that book. It is not clear where these definitions came from: they may have been written by Harriot but are not found in the surviving manuscripts, which leaves open the possibility that they were written by Warner himself.

The relevant Definitions are numbered 14, 15 and 16, and all are concerned with canonical equations. Definition 14 shows how polynomials arise as products of linear factors (with a as the unknown quantity):[52]

Definition 14: Originals of canonical equations

$$\left.\begin{array}{l} a + b \\ a - c \end{array}\right| \begin{array}{l} = aa + ba \\ \quad - ca - bc \end{array}$$

$$\left.\begin{array}{l} a + b \\ a + c \\ a - d \end{array}\right| \begin{array}{l} = aaa + baa + bca \\ \quad + caa - bda \\ \quad - daa - cda - bcd \end{array}$$

If any of the factors on the left is equal to zero then so is the resulting polynomial on the right, and it may therefore be rearranged with all the unknown terms on the left and a given or known term on the right. Definition 15 describes such equations as primary canonicals:[53]

Definition 15: Primary canonicals established by derivation from the originals

$$aa + ba$$
$$- ca = + bc$$

$$aaa + baa + bca$$
$$+ caa - bda$$
$$- daa - cda = + bcd$$

Definition 16 gives examples of secondary canonicals, which arise from primary canonicals when certain relationships between the coefficients cause one (or more) of the powers to vanish:[54]

Definition 16: Secondary canonicals established by reduction from primary canonicals

$$aa = bb \qquad\qquad [b = c]$$

$$aaa - bba$$
$$- bca$$
$$- cca = + bbc$$
$$+ bcc \qquad [d = b + c]$$

Warner rearranged Harriot's Section (d) in accordance with these definitions so that Section 2 of the *Praxis* lists a number of multiplications of the type given in Definition 14,[55] then shows how to form primary canonicals as in Definition 15;[56] Section 3 is concerned with the reduction of primary canonicals to secondary canonicals as in Definition 16;[57] finally Section 4 lists the roots of each equation.[58] Thus material relevant to each equation is found in three or four separate parts of the *Praxis*. This was very different from Harriot's handling of the same material, where each equation was treated just once in a unified way. To make matters worse, Warner was inconsistent in his ordering of the equations from one section to another (see the Appendix) so that it is no easy matter to follow any given equation through the pages of the *Praxis*.

Warner treated Harriot's Sections (*e*) and (*f*) in similarly piecemeal fashion: Section 5 of the *Praxis* contains selected material from Harriot's Section (*e*) but with a good deal of rearrangement, and without the full working out of solutions that Harriot provided. Section 6 of the *Praxis* contains further material from Section (*e*), and much from Section (*f*), but Warner omitted the crucial part *f* δ), on solving quartics lacking a cube term, so the purpose of over thirty examples in the *Praxis* illustrating removal of the cube term is never made clear.

Part II of the *Praxis* contains the kind of numerical examples that are to be found in Harriot's Sections (*a*), (*b*) and (*c*), but only one of the examples given there, and possibly only by coincidence, found its way into the *Praxis*. A detailed comparison between the *Praxis* and Harriot's Sections (*d*), (*e*) and (*f*) can be found in the Appendix.

The *Corrector*

The *Praxis* was almost a travesty of Harriot's *Treatise on equations*, and no-one saw that more clearly than Torporley, who penned a bitter attack on its editors in a piece entitled *Corrector analyticus artis posthumae Thomae Harrioti* ('An analytic correction of the posthumous work of Thomas Harriot').[59] Such a document, by Harriot's closest mathematical colleague, deserves to be read with some care, but it still awaits full translation and in the meantime its readers must grapple with Torporley's tortuous Latin. Torporley's chief complaint was that the editors had so changed Harriot's work that scarcely a trace of the original remained:[60]

> It is impossible for me not to complain passionately, nor to take it badly, that his editors have thus so utterly changed his method, that not only do they not retain his order but scarcely his words.

That which was worthy of greatest praise they have scattered as if the accidental findings of some illiterate.

After three or four pages of complaint, Torporley described in detail what he thought the editors should have included. The work should have begun, he wrote, with the *Operations of arithmetic in letters* immediately after the title:[61]

> The operations of arithmetic in letters, as they were called by Harriot, where his editors begin, under the title. The exemplified forms of the four operations of arithmetic in letters thus clearly not far from there.

The main text should then have begun, according to Torporley, with Harriot's treatment of radicals:[62]

> First a carefully prepared treatise on surds or, as he called them, radical numbers, this indeed not unrelated to the analytic art . . . For if useless to the Exegesis, why did they make mention of it there? If they made mention, it is certain that it is not useless, why therefore have they not described it?

Finally Torporley described Harriot's *Treatise on equations*. His description is so important to the reconstruction of Harriot's text that it is given here in full:[63]

> On the theory of the analytic art itself he was writing the contents in three divisions, the first part of those thus: *On the generation of canonical equations*, 21 sheets on that theme, paginated together under paragraph d), with two short appendices on the multiplication of roots.
>
> The second part moreover under the title: *On solving equations by reduction* has 29 sheets, as paragraph e). Under the same heading, $f\alpha$) 7 sheets; $f\beta$) also 7 sheets; and succeeding these in the numeration of the sheets, $f\gamma$) to sheet f 18 γ) with a short appendix with two lemmas that should not be disparaged, omitted by them. Then $f\delta$) 8 sheets; $f\epsilon$) 4 sheets; $f\zeta$) also 4 sheets. Next, separately, nine sheets containing old reductions recovered by Harriot's method.
>
> But the third part (thus I am not eager for disagreement) he was writing like Viète.[64] *On solving equations in numbers*, and rightly and deservedly. Not nearly all is Viète's in each example. And in the paragraph supposed a),[65] and in 13 sheets are three examples of quadratics, of which the first is his, the other two are Viète's, and five cubics, all Viète's apart from the first. And five quartics of which the fourth is his, the rest Viète's. And these, according to the method of Viète, are all of affirmative equations. The other part of it,[66]

as paragraph *b*) in 12 sheets, has as Viète has, analysis of powers negatively affected: quadratics in *b*1), *b*2), *b*3), cubics in *b*4) to *b*10),
quartics in *b*10), *b*11), *b*12).

The third part of this,[67] as paragraph *c*), has 18 sheets, and treats the
analysis of avulsed powers,[68] as Viète, where there are multiple roots,
and the limits of each are demonstrated. The examples of this are two
quadratics, four cubics, two quartics.

It seems strange that Torporley chose to place sections lettered (*d*), (*e*) and, (*f*)
ahead of those lettered (*a*), (*b*) and (*c*).[69] Apart from this, however, we have
in this extract from the *Corrector* a complete and almost entirely accurate
description of Harriot's *Treatise on equations*.[70] All the sheets survive, and
apart from some minor discrepancies, are as Torporley described them.

The *Summary*

The *Corrector* is already a helpful guide to Harriot's *Treatise on equations*,
but Torporley went further, and collected all the relevant material together.
Using Harriot's own pagination, he listed the sheets that he thought should
have appeared in the posthumous edition of Harriot's algebra, together with
an abbreviated version of the contents of each. Thus Torporley condensed
over 200 of Harriot's sheets into a twenty-page *précis*. His compilation is
untitled but I will refer to it from now on as the *Summary* (see fig. 6).[71] It was
preserved at Sion College (along with the *Congestor* and the *Corrector*) and is
a unique and invaluable document: not only does it bring together details of
manuscript sheets that were later dispersed, but it is now the best guide we
have to what might have been Harriot's original intentions.[72]

The *Summary* is undated but it contains several references to Warner's
edition of the *Praxis*: for example, 'Prob 16 *et* 17 *mutatis signis W et* 18' and
'*omissa W*'.[73] If, as it appears, these references were written in while Torporley
was compiling the *Summary*, he must have worked on it after Warner had
completed the *Praxis*. This is of some significance as it means that Torporley
still had access to Harriot's manuscripts after the *Praxis* was written.

The *Summary* contains the following material:

(i) Operations of arithmetic in letters
The *Summary* begins with the four sheets of *Operations of arithmetic in letters*,
paginated by Torporley as X.1 to X.4 (see fig. 6 and also figs. 3, 4 and 7).

Fig. 6 The first sheet of Torporley's *Summary*, Sion College MS Arc L.40.2/L.40, f. 35.

(ii) Radicals

Torporley listed and summarized over seventy sheets on surds and binomes (numbers of the form $\sqrt{a} \pm \sqrt{b}$);[74] many, if not all of those sheets are still extant, and from Torporley's version it should be possible to identify and collate them, but I have not so far attempted to do so.

(ii) The *Treatise on equations*

Here Torporley set out exactly the material described in the *Corrector*, and in the same order, that is, Sections *(d)*, *(e)* and, *(f)* followed by *(a)*, *(b)* and *(c)*.

III. Harriot's reputation and influence

The *Praxis* was the book on which Harriot's reputation eventually came to rest, but for many years after his death some of his manuscripts also continued to circulate amongst his friends and admirers. In assessing Harriot's influence on seventeenth-century mathematics it is therefore important to consider how widely such material was disseminated, and how well it was understood.

It is clear from a letter that Aylesbury wrote to the Earl of Northumberland after the publication of the *Praxis* that he and Warner had every intention of publishing more of Harriot's work:[75]

> I purpose, God willing, to set forth other peeces of Mr. Harriot, wherein, by reson of my owne incumbrances I must of necessity desire the help of Mr. W.

We know from Samuel Hartlib that by 1639 the mathematician John Pell was also working on some of Harriot's problems with Warner: '[Pell] hase finished those Problemes of Hariot which Warner should have perfected', wrote Hartlib, 'Sir T. A. promised to let them have Harriot's papers but hee did solve them without them.'[76]

Warner died in 1643, but eight years later, in 1651, Pell, Aylesbury and Charles Cavendish were still discussing Harriot's work. By now Pell was teaching in Breda, and Aylesbury and Cavendish were living in exile in Antwerp. Cavendish wrote to Pell:[77]

> S[r]: Th: Aleyburie remembers him to you and desires to knowe if you would be pleased to shew the use of Mr: Hariots doctrine of triangulare numbers; which if you will doe he will send you the originall; I confess I was so farre in love with it that I coppied it out; though I doute I understand it not all;

Cavendish's copy of Harriot's treatise on triangular numbers still survives,[78] but the most important information revealed in his letter to Pell is that Aylesbury was still holding some of Harriot's original manuscripts thirty years after Harriot died, despite the fact that Aylesbury himself was now living in exile and had lost many of his own papers.[79] Harriot's treatise on figurate numbers, his *De numeris triangularibus*, is now in the same volume of papers as Sections (*a*) and (*c*) of the *Treatise on equations*,[80] and both treatises were among the papers discovered at Petworth in 1784. It seems possible, therefore, that Aylesbury was in possession not only of *De numeris triangularibus* but also of the *Treatise on equations* and other papers as late as 1651, and that the papers were returned to Petworth (as Harriot's Will required) either in the remaining years of Aylesbury's life or shortly after his death in 1657.[81]

Since Cavendish, Warner and Pell were all actively interested in Harriot's mathematics we must ask how far any of them discussed his ideas with others. Cavendish is a particularly important figure in this respect, and perhaps more significant than has been recognized as a disseminator of mathematical ideas. It was almost certainly Cavendish who introduced the work of Viète to William Oughtred,[82] and it was also Cavendish who in the late 1630s brought Cavalieri's ideas on indivisibles from France to England.[83] It may reasonably be supposed, therefore, that Cavendish also carried some of Harriot's ideas from England to his mathematical acquaintances on the continent.

This brings us to the controversial question of whether or not Descartes had learned of Harriot's ideas before he wrote *La géométrie*, published in 1637. Descartes had lived in Paris from 1626 to 1628 and afterwards remained in touch with the French mathematical community through Mersenne. We know that Oughtred's *Clavis mathematicae* and Harriot's *Praxis* were known in Paris during the 1630s, for in a letter drafted (probably to Hartlib) in 1635, Pell wrote:[84]

> But what an age is coming on, joy to think of, when our Verulams, Oughtreds and Harriots are read, admired and understood by such transmarines as are likely to perfect what they begun.

Jean Beaugrand, who had edited Viète's *Isagoge* and *Notae priores* in 1631, thought that he detected in *La géométrie* the influence of both Viète and Harriot,[85] but Descartes denied that he had read either.[86] His notation, however, was strikingly similar to Harriot's, especially in his use of lower case letters and yy for y-squared, though he introduced y^3, y^4, etc. for cubes and higher powers.[87] Descartes' objective in *La géométrie* was the algebraic study of curves, but in Book III he demonstrated rules for solving equations, and his method and vertical layout for the removal of a cube term from a quartic were almost identical to those devised by Harriot:[88]

Harriot (here using *e* as his unknown quantity):

$$Et\ fiat\ ..\ eeee + 4.beee + 6.bbee + 4.bbbe +..\ bbbb \equiv +..aaaa$$
$$Et\ deinde\ .\ -4.beee - 12.bbee - 12.bbbe - 4.bbbb \equiv -4\ baaa$$
$$Et\ .\ .\ .\ .\ .\ .\ . -..ffee - 2.ffbe - 4.ffbb \equiv -..ffaa$$
$$Et\ .\ .\ .\ .\ .\ .\ .\ .\ . -..ddde -..bddd \equiv -..ddda$$
$$\equiv +cccc.$$

Hinc reiectis contradictorijs & ordinatis reliquis,

$$fit\ .\ .\ .\ .\ eeee - 6.bbee - 8.bbbe \equiv + .ccce$$
$$-..ffee - 2.ffbe \qquad + 3.bbbb$$
$$-.ddde \qquad + .ffbb$$
$$+ .dddb.\ \text{æquatio}$$

Descartes (for whom *z* is the equivalent of Harriot's *e*):

$$z + \tfrac{1}{2} a \infty x,\ ac\ \text{scribendum}$$

$$z^4 + 2\,a\,z^3 + \tfrac{3}{2}a\,a\,z\,z + \tfrac{1}{2}a^3\,z + \tfrac{1}{16}a^4$$
$$- 2\,a\,z^3 - 3\,a\,a\,z\,z - \tfrac{3}{2}a^3\,z - \tfrac{1}{4}a^4$$
$$+ 2\,a\,a\,z\,z + 2\,a^3\,z + \tfrac{1}{2}a^4$$
$$- c\,c\,z\,z - a\,c\,c\,z - \tfrac{1}{4}a a c c$$
$$- 2\,a^3\,z - a^4$$
$$+ a^4$$

$$z^4 \ast\ + \tfrac{1}{2}a\,a\,z\,z - a^3\,z + \tfrac{5}{16}a^4 \infty\ 0:$$
$$- c\,c\ - a\,c\,c\ - \tfrac{1}{4}a a c c$$

Most significantly, Descartes based his study of polynomial equations on the principle of factorization, the idea so carefully explored in the work of Harriot. Descartes may have discovered such possibilities independently but his contemporaries could be forgiven for thinking that he had some hint of them, even if indirectly, from Harriot.

According to a story later repeated by the English mathematician John Wallis, Cavendish himself thought that Descartes had learned from Harriot. The story relates a discussion between Cavendish and Roberval after the publication of *La géométrie*:[89]

I admire (saith M. Roberval) that notion in Des Cartes of putting over the whole equation to one side, making it equal to Nothing, and how he lighted upon it. The reason why you admire it (saith Sir Charles) is because you are a French-man; for if you were an English-man, you would not admire it. Why so? (saith M. Roberval) Because (saith Sir Charles) we in England know where he had it; namely from Harriot's algebra. What Book is that? (saith M. Roberval,) I never saw it. Next time you come to my Chamber (saith Sir Charles) I will shew it you. Which after a while, he did: And upon perusal of it, M. Roberval exclaimed with admiration (*Il l'a veu! Il l'a veu!*) He had seen it! He had seen it! Finding all that in Harriot which he had before admired in Des Cartes; and not doubting but that Des Cartes had it from thence.

The story was told to Wallis by Pell, who by the late 1630s was acquainted not only with Aylesbury and Warner but with Cavendish himself.

In England Harriot's influence persisted for many years, above all in notation. Almost every mid seventeenth-century English algebraist used Harriot's lower case *a* and *e* for unknowns, though often combined with Descartes' superscript notation for powers.[90] The first book to draw explicitly on Harriot's mathematics was Thomas Gibson's *Syntaxis mathematica* published in 1655. In fact the book combined the work of Harriot and Descartes, as Gibson explained in his preface:[91]

> The method here used is the same as in Master Harriot in some place, that is, in such equations as are proposed in numbers. And as in Des Cartes in some other places, that is, in such equations as are solid [cubic] and not in numbers.

Gibson, like Harriot, used *a* and *e* for unknowns and generally *aaa* (but occasionally a^3) for a cube, and used Harriot's method to solve quadratic and cubic equations numerically. He then went on to show how a polynomial could be written as a product of factors, and stated a number of results concerning the roots and coefficients of polynomial equations. The *Syntaxis* brought together in a concise and readable form the work of both Harriot and Descartes, but for some reason it never became widely known: John Collins, generally well informed about mathematical texts, wrote in 1668 that he had never seen it.[92]

Gibson had borrowed some of Warner's papers from Herbert Thorndike in 1650,[93] but there is no evidence that he saw Harriot's original manuscripts, which at this period were more probably held by Aylesbury. During the early 1650s some of Warner's papers were also lent to Seth Ward, Savilian Professor of Astronomy at Oxford,[94] and to Pell,[95] while others were given to Justinian Isham (and were also seen by Pell).[96] In 1667 some of Warner's papers were lent to Collins, but the accompanying inventory indicates that none of Harriot's algebra was amongst them.[97] In fact Warner's surviving papers contain nothing of significance in relation to Harriot's *Treatise on equations* except a single sheet in Warner's handwriting with material from *d*.4) and *e*.10) on conjugate equations.[98]

Pell's name has now been mentioned several times in connection with Harriot's manuscripts from the late 1630s onwards. Strangely, amongst the hundreds of mathematical papers left by Pell there is hardly anything relating to Harriot except for some scattered notes on the *Praxis*,[99] but it seems that during the early 1670s he did discuss Harriot's algebra with John Wallis, Savilian Professor of Geometry at Oxford.[100] A few years later Wallis published *A treatise of algebra both historical and practical*, and devoted no fewer than twenty-six of its one hundred chapters to Harriot's algebra. Though not published until 1685, *A treatise of algebra* was largely written in the years 1673 to 1676,[101] and Wallis later admitted that his account had been both encouraged and approved by Pell:[102]

> From [Pell's] words I have described what I have said on this matter, and after it was described, I showed it to him (to be examined, changed or emended as he decided or preferred) before it was submitted to the press, and everything that was published was said with Pell's assent and approval.

Pell's influence was not entirely helpful: in *A treatise of algebra* Wallis extolled Harriot to such a degree that it seemed to many, then and since, that his account was grossly exaggerated. Wallis clearly knew the *Praxis* thoroughly,[103] but he also made statements that could not be substantiated from the *Praxis*, for instance that Harriot paid attention to negative as well as positive roots:[104]

> And here first, Beside the *Positive* or *Affirmative* Roots, (which he doth, through his whole Treatise, more especially pursue, as the principal and most considerable:) He takes in also the *Negative* or *Privative* Roots; which by some are neglected.

There are no negative roots in the *Praxis* but they abound in Harriot's *Treatise on equations*, suggesting that the 'Treatise' referred to here by Wallis was not, as is usually supposed, the *Praxis* itself, but the *Treatise on equations* in its original form.

There is further evidence that Wallis knew some of the original manuscript material. As an illustration of Harriot's work on equations, for example, he used the equation $(a - b)(a + c) = 0$ from Harriot's sheet $d.1$), and gave it as a single unified example,[105] though that same material is scattered over three sections of the *Praxis*. Wallis also wrote, as Harriot always did, the algebraic zero oo, though this form does not appear in the *Praxis*. And in a letter to Samuel Morland in 1689, Wallis cited examples of Harriot's negative and imaginary roots: $a = -f$ and $a = \pm \sqrt{-df}$, and gave the page numbers of the *Praxis* where the relevant equations appear.[106] The *Praxis*, however, never gives either negative or imaginary roots, and those quoted by Wallis are to be found only in manuscript, in Sheets $d.7.2$) and $d.13.2$) of the *Treatise on equations*. Thus although Wallis claimed that he no longer knew the whereabouts of Harriot's original papers,[107] he displayed a remarkably intimate knowledge of their contents, a knowledge that can only have come from Pell. Wallis's letter to Morland indicates that he may still have held copies of at least some of the relevant material as late as 1689, four years after Pell had died.

Besides appearing to exaggerate Harriot's achievements, Wallis further harmed his own reputation and Harriot's by reopening the controversy over Descartes' priority. Wallis repeatedly accused Descartes of using Harriot's work without acknowledgement:[108]

> [Harriot] hath laid the foundation on which Des Cartes (though without naming him,) hath built the greatest part (if not the whole) of his Algebra or Geometry. Without which, that whole superstructure of Des Cartes (I doubt) had never been.

The French mathematician Jean Prestet immediately accused Wallis of bringing up old charges without new evidence:[109]

> It is only on vain conjectures or from envy that some have wanted to make believe even in his lifetime that [Descartes] took his method from others, and particularly from a certain English Harriot, whom he had never read, as he declared in one of his letters. And while Monsieur Wallis, a little too jealous of the glory with which France has acquitted herself in mathematics, has just renewed this ridiculous accusation, one is right not to believe it at all, for he speaks without proof.

Wallis's accusations were also taken up by Descartes' biographer, Adrien Baillet, who complained that although Roberval had done all he could to diminish Descartes' reputation, Pell, Aylesbury and Warner had judged Descartes more favourably and were of the opinion that it was honour enough for Harriot that Descartes knew of him.[110] Baillet, whose account is to be admired more for its enthusiasm than its accuracy, did not elaborate on when or where Pell, Aylesbury or Warner had said such things, but goaded by Baillet's remarks, Wallis was forced to defend his position, and to reveal the full extent of Pell's involvement in what he himself had written.[111]

Pell's influence does much to explain Wallis's antagonism towards Descartes. Wallis himself had never communicated with Descartes who had died in 1650, almost at the beginning of Wallis's mathematical career, but Pell had met Descartes in the Netherlands in 1646, and tensions had emerged almost immediately. Descartes apparently discouraged Pell in his efforts to edit Apollonius and Diophantus,[112] and he also reacted with indifference to Pell's *Idea of mathematicks*: 'I inspected the *Mathematical idea* only incidentally', he wrote, 'and now only recollect that there was nothing with which I greatly disagreed.'[113] Pell's dislike of Descartes was well enough known for John Collins to write of it to Leibniz:[114]

> The said Doctor [Pell], being censorious of others, ... was at last censured by Des Cartes for those assertions, concerning whom the Doctor never had any extraordinary esteem.

It seems that Pell's lack of esteem for Descartes communicated itself to Wallis and eventually found its way into the pages of *A treatise of algebra*, where it did immense harm to Wallis's reputation. The most unfortunate consequence of this was that Wallis's opinion of Harriot's algebra was never taken seriously,[115] yet in many respects his account was both accurate and perceptive. As a mathematician himself, Wallis saw clearly what Harriot had achieved:[116]

> Beside those convenience in the Notation ... Mr. Harriot, as to the *Nature of Equations*, (wherein lyes the main Mystery of *Algebra*;) hath made much more improvement. Discovering the true Rise of Compound Equations; and Reducing them to the Originals from whence they arise.

A century later, in 1796, the same views were repeated by Charles Hutton:[117]

> [Harriot] shewed the universal generation of all the compound of affected equations, by the continual multiplication of so many simple ones, or binomial roots; thereby plainly exhibiting to the eye the whole circumstances of the nature, mystery and number of the roots of equations; with the composition and relations of the coefficients of the terms; and from which many of the most important properties have since been deduced.

By the time Hutton was writing, Harriot's papers had at last resurfaced, having been found at Petworth in 1784 by Franz Xavier von Zach, and their reappearance led to a reappraisal of Harriot's algebra and astronomy. In the early 1790s Oxford University Press undertook to publish some of the newly discovered manuscripts, and Hutton wrote optimistically:[118]

> ... it is with pleasure I can announce, that they are in a fair train to be published: they have been presented to the university of Oxford, on condition of their printing them; with a view to which, they have been lately put into the hands of an ingenious member of that learned body, to arrange and prepare them for the press.

Abraham Robertson, Savilian Professor first of geometry (1797–1810) then of astronomy (1810–26), was that 'ingenious member', who inspected the sheets of Section (*b*) but could make little of them:[119]

> Thirteen of these [sheets] are entitled '*De numerosa potestatum res*' and contain problems similar to those in the '*Exegisimus*' of the *Praxis*. Among these, several are marked as Vieta's, and this shews how necessary it is to be cautious in reasoning on what belongs to different individuals.

Robertson failed to realize that Harriot, far from plagiarizing Viète's work, was in fact re-writing it in his own much clearer notation, noting carefully which examples were taken from Viète.

Robertson's lack of enthusiasm and Zach's failure to deliver the promised introduction meant that by 1799 the University Press had no choice but to cancel the contract, and the manuscripts were returned to Petworth.[120] Unfortunately, in the course of Zach's investigations (if not earlier) the sections of Harriot's *Treatise on equations* had become separated, so that it was no longer easy to recognize a unified treatment in the disordered sheets. There was further fragmentation when the papers were divided between the British Library and Petworth; only Section (*b*) remained at Petworth, and from then on any sense of a unified treatise was lost. Stephen Rigaud, Robertson's successor in both Savilian chairs, took a keen interest in Harriot and his work,[121] and even published a sheet of Harriot's algebra in his *Supplement to Dr Bradley's miscellaneous works*,[122] but after his death in 1839, interest in Harriot and his mathematics faded until the second half of the twentieth century.

By then Harriot as an algebraist had almost disappeared from view. Invaluable and sustained research on Harriot's scientific papers was done by Johannes Lohne, but in his comprehensive review of Harriot's mathematical and scientific writing he had little to say about Harriot's work on equations: he remarked, in a footnote only, on sheets marked '(*e*.1) to (*e*.29) and sometimes (*f*)' but failed to recognize them as part of a single unified treatise.[123] Only Cecilia Tanner, who did so much to bring about the revival of Harriot studies, recognized that Torporley's *Summary* was the clue to reconstructing Harriot's work on equations, but took her discovery no further than providing a list of the relevant manuscript sheets.[124] Meanwhile most modern historians of mathematics, even specialist historians of algebra, have failed to mention Harriot,[125] while others have based their judgements on the material in the *Praxis*,[126] and so have been unable to recognize the true extent and nature of his insights.

I am delighted that 200 years after the papers were rediscovered at Petworth, Oxford University Press is at last able to publish some of Harriot's mathematics. This edition offers, for the first time since 1632, a complete version of Harriot's *Treatise on equations*, and in doing so will perhaps begin to fulfil the hopes held long ago by Lower, Torporley, Pell, Wallis and others: that Harriot should be established in his rightful place in the history of the evolution of algebra.

A note on conventions adopted in preparing the text

I have adhered as closely as possible to Harriot's original notation and layout. However, as in any translation, it has sometimes been necessary to sacrifice literal renderings for the sake of clarity and readability. Harriot's sheets are often closely written using two columns, and in general I have spaced these over two separate pages. I have also made changes where Harriot wrote two, three or more equations linked together by either horizontal or vertical equality signs; in such cases I have separated out the individual equations, and have written them horizontally, in accordance with modern conventions. Where I have made occasional small insertions to improve the readability of the text, these are shown in square brackets.

In the *Operations of arithmetic in letters* I have retained Harriot's notation exactly, but thereafter I have dropped his use of commas for multiplication (as he himself often did), writing simply $4xxz$, etc. Harriot also used commas to indicate parenthesis; I have replaced these in the few instances where they are needed with brackets.

Treatise on equations

2.)

multipl. a.
in. b.
facta. ab.

| $aa.$
$bb.$
$bbaa.$
vel. baba.

| bc
d
$bcd.$

| bbb
bb
$bbbbb$

| $bbcc$
dd
$bbccdd.$
vel: bcdbcd.

multip. b+a.
in. b+a.
$bb+ba+aa$
$+ba+aa$
facta. $bb+2ba+aa.$

| $b−a$
$b−a$
$bb−ba$
$−ba+aa$
$bb−2ba+aa.$
vel: $bb−,2ba+aa.$

| $b+a$
$b−a.$
$bb+ba$
$−ba−aa$
$bb−aa.$

multip. b+c+d.
in. a.
facta. $ba+ca+da$

| $b+c−d.$
$b−c+d$
$bb+bc−bd$
$−bc−cc+cd$
$+bd+cd−dd$
$bb−cc+2cd−dd.$

| $8−2$
$8−2$
$64−16$
$−16+4$
$68−32$
36

Applica. bc
ad. a
orta. $\dfrac{bc}{a}$

| aa
b
$\dfrac{aa}{b}$

| bbb
ca
$\dfrac{bbb}{ca}$

| $bbdd$
cd
$\dfrac{bbdd}{cd}$ | $=\dfrac{bbd}{c}$

Applica. $bbcc$
ad cc
orta. $bb.$

| $bcdf.$
$bdf.$
$c.$

| $bcdf$
cf
bd

Applica. $ba+ca+da$
ad. a.
orta. $b+c+d.$

| $ba+ca+da$
$b+c+d$
$a.$

$\dfrac{bb+2ba+aa}{b+a}=b+a$ $\dfrac{bb−aa}{b−a}=b+a$

$\dfrac{bbb+ccc}{b+c}=bb−bc+cc.$ $\dfrac{bbb−ccc}{b−c}=bb+bc+cc.$

manifestū
per praecog=
nitū genera=
tionē—.

Fig. 7   Sheet 2) of *Operations of arithmetic in letters*, Add MS 6784, f. 323.

37

Operations of arithmetic in letters

1)   *Operations of arithmetic in letters*[1]

	a			aa			aaa
add	b		add	bc		add	bcc
sum	$a+b$		sum	$aa+bc$		sum	$aaa+bcc$

	a			aa			aaa
subtract	b		subtract	bc		subtract	bcc
remainder	$a-b$		remainder	$aa-bc$		remainder	$aaa-bcc$

	$a+b$		$a+b$		$a+b$	$a+b$
add	$c+d$		$c-d$		$-d$	$-b$
sum	$a+b+c+d$		$a+b+c-d$		$a+b-d$	a

	$a+b$		$aa+cc$		$aaa+cdf-ddd$
add	$c+b$		$aa+cc$		$aaa+bdd+ddd$
sum	$a+c+2{,}b$		$2{,}aa+2{,}cc$		$2{,}aaa+cdf+bdd$

	$a+b$		$a+b$		$a+b$	$a+b$
subtract	$c+d$		$c-d$		$-d$	$-b$
remainder	$a+b-c-d$		$a+b-c+d$		$a+b+d$	$a+2{,}b$
			or $a+b-{,}c-d$			

	$a+b$		$aa+cc$		$aaa+cdf-ddd$
subtract	$c+b$		$aa+cc$		$aaa+bdd+ddd$
remainder	$a+b-c-b$		$aa+cc-aa-cc$		$cdf-bdd-2{,}ddd$
that is:	$a-c$		that is: 0		

$b + 7,a$	$b + 7,a$	$b + 9,a$	$b - 9,a$
add $\quad +9,a$	$-9,a$	$b - 7,a$	$b + 7,a$
sum $\quad b + 16,a$	$b - 2,a$	$2,b + 2,a$	$2,b - 2,a$

$b + 7,a$	$b + 7,a$	$b + 9,a$	$b - 9,a$
subtract $\quad +9,a$	$-9,a$	$b - 7,a$	$b + 7,a$
remainder $\quad b - 2,a$	$b + 16,a$	$16,a$	$-16,a$

Add MS 6784 f. 323

2)

multiply	a	aa	bc	bbb	$bbcc$
by	b	bb	d	bb	dd
product	ab	$bbaa$	bcd	$bbbbb$	$bbccdd$

multiply	$b + a$	multiply	$b - a$	$b + a$
by	$b + a$	by	$b - a$	$b - a$
	$bb + ba$		$bb - ba$	$bb + ba$
	$+ ba + aa$		$- ba + aa$	$- bb - aa$
product	$bb + 2ba + aa$	product	$bb - 2ba + aa$	$bb - aa$
		or	$bb - ,2ba - aa$	

multiply	$b + c + d$	$b + c - d$	$8 - 2|$
by	a	$b - c + d$	$8 - 2|$
product	$ba + ca + da$	$bb + bc - bd$	$64 - 16$
		$- bc - cc + cd$	$- 16 + 4$
		$+ bd + cd - dd$	$68 - 32$
		$bb - cc + 2,cd - dd$	$= 36$

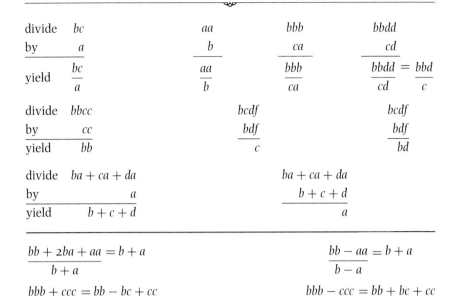

divide	bc		aa	bbb	$bbdd$
by	a		b	ca	cd
yield	$\dfrac{bc}{a}$		$\dfrac{aa}{b}$	$\dfrac{bbb}{ca}$	$\dfrac{bbdd}{cd} = \dfrac{bbd}{c}$

divide	$bbcc$		$bcdf$		$bcdf$
by	cc		bdf		bdf
yield	bb		c		bd

divide	$ba + ca + da$		$ba + ca + da$
by	a		$b + c + d$
yield	$b + c + d$		a

$$\frac{bb + 2ba + aa = b + a}{b + a} \qquad \frac{bb - aa = b + a}{b - a}$$

$$\frac{bbb + ccc = bb - bc + cc}{b + c} \qquad \frac{bbb - ccc = bb + bc + cc}{b - c}$$

Evident from the previously investigated forms.

Add MS 6784 f. 324

3)

$$\frac{ba = a}{b} \qquad\qquad \frac{bca = ca}{b} \qquad\qquad \frac{bca = ba}{c}$$

$$\frac{ba}{c} + z = \frac{ba}{c} + \frac{zc}{c} = \frac{ba + zc}{c}$$

$$\frac{ae}{b} + z = \frac{ae + zb}{b}$$

$$\frac{ac}{b} + \frac{zz}{g} = \frac{acg}{bg} + \frac{bzz}{bg} = \frac{acg + bzz}{bg}$$

$$\frac{ac - z}{b} = \frac{ac}{b} - \frac{zb}{b} = \frac{ac - zb}{b}$$

$$\frac{ac - zz}{b} = \frac{acg}{g} - \frac{zzb}{bg} = \frac{acg - zzb}{bg}$$

$$\left. \begin{array}{l} \dfrac{ac}{b} \\ \dfrac{}{b} \\ b \end{array} \right| = \dfrac{acb}{b} = ac \quad \left. \begin{array}{c} \dfrac{ac}{b} \\ \\ \dfrac{zz}{g} \end{array} \right| = \dfrac{aczz}{bg}$$

$$\left. \begin{array}{l} \dfrac{ac}{b} \\ z \end{array} \right| = \dfrac{acz}{b}$$

$$\left. \begin{array}{l} \dfrac{aaa}{b} \\ d \end{array} \right| = \dfrac{aaa}{bd} \qquad \text{or} \qquad \left. \begin{array}{l} \dfrac{aaa}{b} \\ d \end{array} \right| \left. \begin{array}{l} = \dfrac{aaa}{b} \\ \dfrac{db}{b} \end{array} \right| = \dfrac{aaa}{db}$$

$$\left. \begin{array}{l} bg \\ \dfrac{ac}{d} \end{array} \right| = \dfrac{bgd}{ac} \qquad \text{or} \qquad \left. \begin{array}{l} bg \\ \dfrac{ac}{d} \end{array} \right| \left. \begin{array}{l} = \dfrac{bgd}{d} \\ \dfrac{ac}{d} \end{array} \right| = \dfrac{bgd}{ac}$$

$$\left. \begin{array}{l} bbb \\ \dfrac{z}{} \\ \dfrac{aaa}{dc} \end{array} \right| = \dfrac{bbbdc}{zaaa} \qquad \text{or} \qquad \left. \begin{array}{l} bbb \\ \dfrac{z}{} \\ \dfrac{aaa}{dc} \end{array} \right| \left. \begin{array}{l} = \dfrac{bbbdc}{zdc} \\ \dfrac{aaaz}{dcz} \end{array} \right| = \dfrac{bbbdc}{zaaa}$$

4)

Suppose	$aa - dc = gg - ba$	
I say that	$aa + ba = gg + dc$	by antithesis
Because	$aa - dc = gg - ba$	
add to each side	$ba + dc \quad ba + dc$	
then	$aa + ba = gg + dc$	
Second, suppose	$aa - dc = gg$	
I say that	$aa = gg + dc$	
Because	$aa - dc = gg$	
add to each side	$+dc \quad\quad + dc$	
then	$aa = gg + dc$	
and thus	$aa - gg = dc$	

Suppose	$aaa + baa = zca$	
I say that	$aa + ba = zc$	by hypobibasmus

Suppose	$baa + dca = zcd$	
I say that	$aa + \dfrac{dca}{b} = \dfrac{zcd}{b}$	by parabolismus
Or suppose	$baa + dba = zbd$	
I say that	$aa + da = zd$	

Fig. 8   Sheet [*a*].8) of the *Treatise on equations*, Add MS 6782, f. 392.

On solving equations in numbers

[a].1)   *On solving equations in numbers*[1]

Problem 1) $aa + da = xz$ Viète folio 7.b

$$aa + 24a = 2356$$

$$38 + 24 = 2356$$

$$38 \quad 38$$

Canonical form:

$$\left. \begin{array}{l} b+c \\ \overline{b+c} \end{array} \right| + \left. \begin{array}{l} d \\ \overline{b+c} \end{array} \right| = 2356$$

that is:

$$\left. \begin{array}{l} bb \\ +\, bd \end{array} \right| + 2bc + cc = 2356$$
$$ + dc$$

or:

$$\left. \begin{array}{l} b+d \\ b \end{array} \right| + \left. \begin{array}{l} 2b+d \\ c \end{array} \right| + \left. \begin{array}{l} c \\ c \end{array} \right| = 2356$$

Solution according to Viète's method, a little changed.

		b	c	
		4		
	2	$\dot{3}$	5	$\dot{6}$
		2	$\dot{4}$	
bb		1	6	
bd			9	6
$bb + bd$	2	5	6	

45

It is therefore not possible to carry out the intermediate operation and clearly *b* must be 3.

			b		*c*
			3		8̇
		2	3̇	5	6̇
			2	4̇	.
bb			9		
bd			7	2	
bb + *bd*		I	6	2	
2*bc* + *cd* + *cc*			7̇	3	6̇	*c* = 8
d				2	4
2*b*				6	
2*b* + *d*				8	4
dc			I	9	2
2*bc*			4	8	
cc				6	4
2*bc* + *cd* + *cc*			7	3	6
			O	O	O

Another way:

				4			
			2	3̇	5	6̇	
d				2	4̇	.	
b				4			
b + *d*				6	4		
bd				9	6		
X.	*bb* + *bd*		2	5	6	X.	

		b		c
		3		8
	2	$\dot{3}$	5	$\dot{6}$
d		2	$\dot{4}$	
b		3		
b + d		5	4	
bb + bd	I	6	2	
2bc + cd + cc		$\dot{7}$	3	$\dot{6}$
2b		6		
d			2	4
2b + d		8	4	
2b + d		6	7	2
cc			6	4
2bc + dc + cc		7	3	6
		0	0	0

$c = 8$

Add MS 6782 f. 398

[a].2)   *On solving equations in numbers*

Problem 1) $aa + da = xz$

$$243 + 7 = 60750$$
$$243 \quad 243$$
$$aa + 7a = 60750$$

Viète's first example[2]

Canonical form
as above:

$$bb \mid\; + 2bc + cc = 60750$$
$$bd \mid\; + dc$$

that is:

$$b+d \mid\; + 2b+d \mid\; + c \mid\; = 60750$$
$$b \mid\; c \mid\; c \mid$$

47

Viète's solution, a little changed.

	c1	c2	c3	c4	c5	
			b		c	
		b	c			
		2	4		3	
	6̇	0	7̇	5	0̇	
d			7̇	.	.	
bb	4					
db			1	4		
bb + db	4		1	4		
2bc + dc + cc		1	9	3	5	*c* = 4
2b + d			4	0	7	
2bc + dc		1	6	2	8	
cc			1	6		
2bc + dc + cc		1	7	8	8	
2bc + dc + cc		1	4	7	0	*c* = 3
d					7	
2b			4	8		
2b + d			4	8	7	
cd				2	1	
2b		1	4	4		
cc					9	
2bc + cd + cc		1	4	7	0	
		0	0	0	0	

[*a*].3)   *On solving equations in numbers*

Problem I) $aa + da = xz$

$28 + 762 = 22120$

$28 \quad\ \ 28$

$aa + 762a = 22120$ The second case, by division

Canonical form $bb \mid + 2bc + cc = 22120$
as above: $bd \mid + dc$

or: $b + d \mid + 2b + d \mid + c \mid = 22120$
                     $\quad\ \ b \mid \qquad\ \ c \mid\ \ c \mid$

Solution.

$$\begin{array}{ccccc} \dot 2 & 2 & \dot 1 & 2 & \dot 0 \\ 7 & 6 & \dot 2 & \cdot & \cdot \end{array}$$

Since $7 > 2$ it is a case of devolution. (Viète's example is $aa + 954a = 18487$.)[3]

			b	*c*	
			2	8	
	2	2	$\dot 1$	2	$\dot 0$
d		7	6	$\dot 2$	$\cdot$
b			2		
b + *d*		7	8	2	
bb + *bd*	I	5	6	4	
2*bc* + *dc* + *cc*		6	$\dot 4$	8	$\dot 0$	*c* = 8
d			7	6	2
2*b*				4	
2*b* + *d*			$\dot 8$	0	$\dot 2$
2*bc* + *dc*		6	4	I	6
cc				6	4
2*bc* + *dc* + *cc*		6	4	8	0
		0	0	0	0

[a].4)   *On solving equations in numbers*

Problem 2) $aaa + dda = xxz$

$$28 + \quad 35 = 22932$$
$$28 \qquad \text{I}$$
$$28 \qquad 28$$

$$aaa + 35a = 22932$$

Canonical form:

$$\begin{aligned} b + c + \quad d \\ b + c \qquad d \\ b + c \quad b + c \end{aligned} \Big| = 22932$$

that is:

$$\left.\begin{aligned} bbb \\ + dbb \end{aligned}\right| \begin{aligned} +3bbc + 3bcc + ccc = 22392 \\ + ddc \end{aligned}$$

or:

$$bb + bd \Big|{}_{b} +3bb + dd \Big|{}_{c} + 3b \Big|{}_{c}^{c} + c \Big|{}_{c}^{c} = 22932$$

Solution.

		b			c	
		2			8	
	2	$\dot{2}$	9	3	$\dot{2}$	
dd				3	$\dot{5}$	
bbb		8				
ddb			7	0		
$bbb + ddb$		8	7	0		
$3bbc + 3bbc + ccc + ddc$	I	$\dot{4}$	2	3	$\dot{2}$	$c = 8$
dd				3	5	
$3bb$		I	2			
$3b$				6		
$3bb + dd + 3b$		I	2	9	5	
$3bbc$		$\dot{9}$	6			
$3bcc$		3	8	4		
ccc			5	I	2	
ddc			2	8	0	
$ddc + 3bbc + 3bbc + ccc$	I	4	2	3	2	
	0	0	0	0	0	

Add MS 6782 f. 395

[a].5) *On solving equations in numbers*

Problem 2) $aaa + dda = xxz$

$243 + 30 = 14356197$

$243 \quad$ I

$243 \quad 243$

$aaa + 30a = 14356197$ Viète's first example[4]

Canonical form: $\quad bbb \big| + 3bbc + 3bcc + ccc$
$\quad\quad\quad\quad\quad + ddb \big| + ddc$

or: $\quad bb + dd \big| + 3bb + dd \big| + 3b \big| + c \big|$
$\quad\quad\quad\quad\quad b \big| \quad\quad\quad\quad c \big| \quad\quad c \big| \quad c \big|$
$\quad\quad\quad\quad\quad\quad\quad\quad\quad\quad\quad\quad\quad\quad c \big| \quad c \big|$

Solution.

			b			c				
			b			c				
			2			4			3	
	1	4̇	3	5	6̇	1	9	7̇		
dd					3	Ȯ				
bbb		8								
ddb					6	0				
bbb + ddb		8̇	0	0	6̇	0				
ddc + 3bbc + 3bcc + ccc		6̇	3	5	0̇	1	9			c = 4
dd						3	0			
3bb		1	2							
3b				6						
dd + 3bb + 3b		1̇	2	6	0̇	3	0			
ddc					1	2	0			
3bbc		4	8							
3bcc			9	6						
ccc				6	4					
ddc + 3bbc + 3bcc + ccc		5̇	8	2	5̇	2	0			
		5	2	4	9	9	7			

Now b becomes 24

		5	2	4	9	9	7			c = 3
dd						3	0			
3bb			1	7	2	8				
3b						7	2			
dd + 3bb + 3b			1	7	3̇	5	5	Ȯ		
ddc							9	0		
3bbc			5	1	8	4				
3bcc					6	4	8			
ccc							2	7		
ddc + 3bbc + 3bcc + ccc			5	2	4	9	9	7		
			0	0	0	0	0	0		

[a].6)   *On solving equations in numbers*

Problem 2) $aaa + dda = xxz$

$$19 + 95400 = 1819459$$

19	I
19	19

$aaa + 95400a = 1819459$ Viète's second example [5]

Canonical form: $ddb \;|\; + ddc$
                     $+ bbb \;|\; + 3bbc + 3bcc + ccc$

Solution.   by division

				b		c	
		O		I		9	
		i̇ 8	I	9̇	4	5	9̇
dd		9	5	4	ȯ	ȯ	·
ddb		9	5	4̇	0	0	
bbb				I			
$ddb + bbb$		9	5	5̇	0	0	
$ddc + 3bbc + 3bcc + ccc$		8	6	4̇	4	5	9̇
dd		9	5	4	0	0	
$3bb$				3			
$3b$					3		
$dd + 3bb + 3b$		9	5̇	7	3	ȯ	
ddc		8	5	8̇	6	0	ȯ
$3bbc$			2	7			
$3bcc$			2	4	3		
ccc				7	2	9	
$ddc + 3bbc + 3bcc + ccc$		8	6	4̇	4	5	9̇
		0	0	0	0	0	0

$c = 9$

Add MS 6782 f. 393

[a].7)   *On solving equations in numbers*

Problem 3) $aaa + daa = xxz$

$$432 + 30 = 86220288$$
$$432 \qquad 432$$
$$432 \qquad 432$$

$aaa + 30a = 86220288$ Viète's first example[6]

Canonical form:

$$
\begin{array}{llll}
b + c + & d & = & bbb \mid + 3bbc + 3bcc + ccc \\
b + c & b + c & & + bbd \mid + 2bdc + dcc \\
b + c & b + c & &
\end{array}
$$

Solution.

		b			c			
		4			3			
	8	$\dot{6}$	2	2	$\dot{o}$	2	8	$\dot{8}$
d					3	$\dot{o}$		
bbd			4	8	0			
bbb		6	4					
$bbd + bbb$		6	8	8	0			
	1	$\dot{7}$	4	2	$\dot{o}$	2	$c = 3$	
d						3	$\dot{o}$	
$2bd$				2	4	0		
$3bb$			4	8				
$3b$				1	2			
$d + 2bd + 3bb + 3b$			5	1	6	3	0	
dcc				2	7	0		
$2bdc$				7	2	0		
$3bbc$			1	4	4			
$3bcc$			1	0	8			
ccc				2	7			
$dcc + 2bdc + 3bbc$ etc.		1	$\dot{6}$	2	5	$\dot{4}$	0	
		1	1	6	6	2	8	8

Now *b* becomes 43

	1	1	6	6̇	2	8	8̇		*c* = 2
d					˙	3	0̇		
2*bd*			2	5	8	0			
3*bb*		5	5	4	7				
3*b*				1	2	9			
d + 2*bd* etc.		5	8	1̇	8	2	0̇		
B *dcc*				1	2	0			
2*bdc*			5	1	6	0			
3*bbc*	1	1	0	9	4				
3*bcc*			5	1	6				
ccc							8		
B	1	1	6	6	2	8	8		
	0	0	0	0	0	0	0		

Add MS 6782 f. 392

[*a*].8)   *On solving equations in numbers*

Problem 3)   *aaa* + *daa* = *xxz*

---

24 + 10000 = 5773824

24   24

24   24   Viète's second example[7]

aaa + 10000*aa* = 5773824   done by division

Canonical form: *bbb* | + 3*bbc* + 3*bcc* + *ccc*

               + *bbd* | + 2*bdc* + *dcc*

Solution.

Note: the first figure is acquired by division as if the canonical form were:

d + *b* | which is equal to *bbd* + *bbb*

   bb |

that is, if *d* + *b* were the divisor, the quotient would be *bb*, and the root *b* would be the first figure. As here the quotient is 5, whose square root, 2, is put for the first figure. See Viète.

				b		c		
				2		4		
	5̇	7	7	3̇	8	2	4̇	
d	1	0	0̇	0	0̇			
bbd	4	0	0	0	0			
bbb					8			
bbd + bbb	4	0	0	8	0			
2bdc + dcc etc.	1	7	6	5̇	8	2	4̇	c = 4
d			1	0	0̇	0	0̇	
2bd		4	0	0	0	0		
3bb				1	2			
3b						6		
d + 2bd + 3bb + 3b		4	1	1	2	6	0	
B dcc		1	6	0	0	0	0	
2bdc	1	6	0	0	0			
3bbc				4	8			
3bcc				9	6			
ccc						6	4	
B	1	7	6	5	8	2	4	
	0	0	0	0	0	0	0	

[a].9)   *On solving equations in numbers*

Problem 4) $aaaa + daaa = xxxz$

$24 + 1000 = 3557776$

24	I
24	I
24	24

$aaaa + 1000a = 3557776$ Viète's first example[8]

Canonical form:
$$bbbb \mid + 4bbbc + 6bbcc + 4bccc + cccc$$
$$+ dddb \mid + dddc$$

Solution.

			b				c		
			2				4		
		3	5̇	5	7	7	6̇		
	ddd		I	O	O	Ȯ			·
	dddb		2	O	O	O			
	bbbb	I	6						
	dddb + bbbb	I	8	O	O	O			
	dddc + 4bbbc etc.	I	7̇	5	7	7	6̇		c = 4
	ddd			I	O	O	Ȯ		
	4bbb			3	2				
	6bb				2	4			
	4b					8			
	ddd + 4bbb etc.		3	5	4	8	O		
B	dddc			4	O	O	Ȯ		
	4bbbc		I	2	8				
	6bbcc			3	8	4			
	4bccc				5	I	2		
	cccc					2	5	6	
B		I	7̇	5	7	7	6̇		
		O	O	O	O	O	O		

[a].10)   *On solving equations in numbers*

Problem 4) $aaaa + daaa = xxxz$

$$24 + 100000 = 2731776$$

24	I
24	I
24	24

$aaaa + 100000a = 2731776$ Viète's second example[9]
                                    done by division

| Canonical form: | $bbbb$ | $+ 4bbbc + 6bbcc + 4bccc + cccc$ |
| | $+ dddb$ | $+ dddc$ |

Solution.

				b			c
				2			[4]
		2	7	$\dot{3}$	I	7	7	$\dot{6}$
ddd		I	0	0	0	0	$\dot{0}$
$dddb$		2	0	0	0	0	0
$bbbb$			I	6			
$dddb + \overline{bbbb}$		2	I	6	0	0	0
$dddc + 4bbbc$ etc.			5	$\dot{7}$	I	7	7	$\dot{6}$	$c = 4$
ddd			I	$\dot{0}$	0	0	0	$\dot{0}$
$4bbb$				3	2		
$6bb$				2	4		
$4b$						8	
$ddd + 4bbb$ etc.			I	3	4	4	8	0
B	$dddc$		4	$\dot{0}$	0	0	0	$\dot{0}$
	$4bbbc$		I	2	8		
	$6bbcc$			3	8	4	
	$4bccc$				5	I	2
	$cccc$				2	5	6
B			5	$\dot{7}$	I	7	7	$\dot{6}$
			0	0	0	0	0	0

[*a*].11)   *On solving equations in numbers*

Problem 5) $aaaa + daaa = xxxz$

$$24 + 10 = 470016$$
$$24 \quad 24$$
$$24 \quad 24$$
$$24 \quad 24$$

$aaaa + 10aaa = 470016$ Viète's only example[10]

Canonical form:
$$
\begin{array}{llll}
b + c + & d & = & bbbb \\
b + c & b + c & + bbbd \\
b + c & b + c \\
b + c & b + c
\end{array}
\quad
\begin{array}{l}
+\, 4bbbc + 6bbcc + 4bccc + cccc \\
+\, 3bbdc + 3bdcc + dccc
\end{array}
$$

59

Solution.

			b				c	
			2				4	
		4	7̇	0	0	I	6̇	
	d		I	Ȯ				
	bbbd		8	0				
	bbbb	I	6					
	bbbd + bbbb	2	4	0				
	3bbdc + 3bdcc etc.	2	3̇	0	0	I	6̇	$c = 4$
B	d					I	Ȯ	
	3bd				6	0		
	3bbd			I	2	0		
	4bbb			3	2			
	6bb				2	4		
	4b						8	
B			4	7	0	9	0	
C	dccc				6	4	Ȯ	
	3bdcc				9	6	0	
	3bbdc			4	8	0		
	4bbbc	I	2	8				
	6bbcc			3	8	4		
	4bccc				5	I	2	
	cccc				2	5	6	
C		2	3̇	0	0	I	6̇	
		0	0	0	0	0	0	

Another example, of mine, which is done by means of division:

$$aaa + 300aa = 4478976$$

Solution.

For the first figure divide 44 by 3. The quotient is 14, which contains the greatest cube 8, with root 2. This is to be put for the first figure if the rest are consistent.

				b				c		
				2				4		
		4	4	7̇	8	9	7	6̇		
	d			3	0	0̇				
	bbbd		2	4	0	0				
	bbbb			1	6					
bbbd + bbbb			2	5	6̇	0				
			1	9	1̇	8	9	7	6̇	
B	d					3	0	0		
	3bd				1	8	0	0		
	3bbd			3	6	0	0			
	4bbb				3̇	2				
	6bb					2	4			
	4b							8		
B				4	1̇	2	7	8	0̇	
C	dccc				1̇	9	2	0	0̇	
	3bdcc			2	8	8	0	0		
	3bbdc		1	4	4	0	0			
	4bbbc				1	2	8			
	6bbcc					3	8	4		
	4bccc						5	1	2	
	cccc							2	5	6
C			1	9	1̇	8	9	7	6̇	
		0	0	0	0	0	0	0		

$c = 4$

[a].12)   *On solving equations in numbers*

Problem 6) $aaaa + ddaa + fffa = xxxz$

$$24 + 200 + 100 = 449376$$

24	I	I
24	24	I
24	24	24

$aaaa + 200aa + 100a = 449376$ Viète's only example[11]

Canonical form:

$b+c$ +	d	+ f	=	$bbbb$	+ $4bbbc$ + $6bbcc$ + $4bccc$ + $cccc$
$b+c$	d	f		$bbdd$	+ $2bddc$ + $ddcc$
$b+c$	$b+c$	f		$bfff$	+ $fffc$
$b+c$	$b+c$	$b+c$			

Solution.

			b				c	
			2				4	
		4	4̇	9	3	7	6̇	
	fff				1	0	0̇	
	dd				2	0	0̇	
A	*fffb*			·	2	0	0	
	bbdd			8	0	0		
	bbbb	1	6					
A		2	4̇	2	0	0		
		2	0̇	7	3	7	6̇	c = 4
B	*fff*				1	0̇	0̇	
	dd				2̇	0	0̇	
	2bdd				8	0	0	
	4bbb			3̇	2			
	6bb				2	4		
	4b						8	
B			4̇	2	7	8	0̇	
C	*fffc*				4	0	0	
	ddcc			3	2	0	0	
	2bddc		3	2	0	0		
	4bbbc	1	2	8				
	6bbcc			3	8	4		
	4bccc			5	1	2		
	cccc				2	5	6	
C		2	0	7	3	7	6	
		0	0	0	0	0	0	

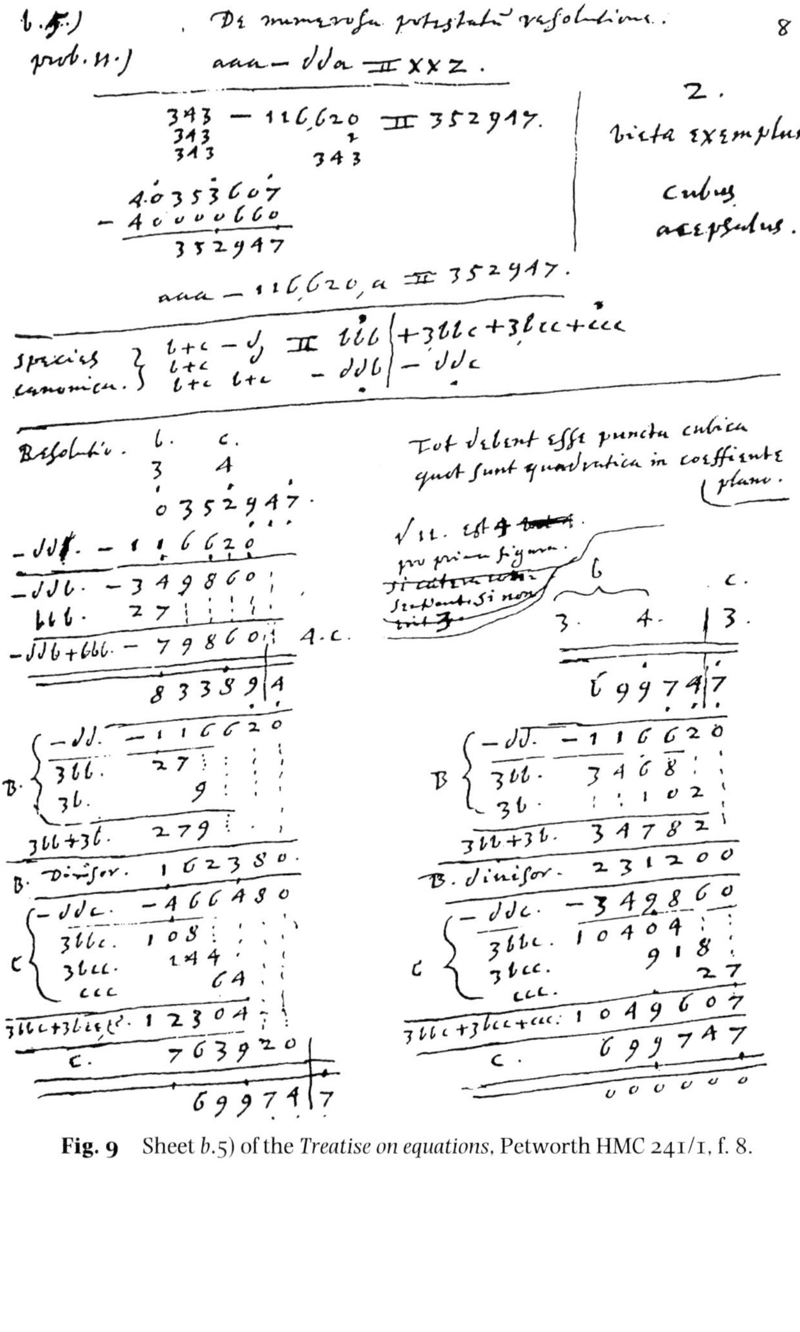

Fig. 9 Sheet *b*.5) of the *Treatise on equations*, Petworth HMC 241/1, f. 8.

On solving equations in numbers

b.1)   *On solving equations in numbers*

Problem   10) $aa - da = xz$

$$250 - 7 = 60750$$
$$250 \quad 250$$

$aa - 7a = 60750$ Viète's first example[1]

Canonical form: $\begin{array}{ll} b+c-d & = \\ b+c \quad b+c & \end{array} \left. \begin{array}{l} bb \\ -bd \end{array} \right| \begin{array}{l} +2bc + cc \\ -dc \end{array}$

or: $\left. \begin{array}{c} b-d \\ b \end{array} \right| \begin{array}{c} +2b - d + cc \\ c \end{array}$

Solution.

The quadratic points must be the same in number as the single places after the coefficient of the linear term.

```
                          b           c
                  b       c
                  2       5       0
                  6̇   0   7̇   5   0̇
        − d                   7̇
        bb        4
        − bd          −   1   4
     bb − bd       3   8   6
                      2̇   2   i   5       c = 5
        − d                  −   7̇
        2b                4
     2b − d           3   9   3       divisor
        − dc          −   3   5̇
        2bc       2̇   0
        cc                2   5
     2bc + cc     2   2   5
  2bc + cc − dc   2̇   2   i   5
                  0   0   0̇   0̇       c = 0
```

HMC 241/1 f. 12

b.2) On solving equations in numbers

Problem   10) $aa - da = xz$

$242 - 242 = 484$

250   250

$aa - 240a = 484$ Viète's second example[2]
                                 acephalic square[3]

Canonical form:

$$b + c - d \quad = \quad bb \;\big|\; + 2bc + cc$$
$$b + c \quad b + c \quad - bd \;\big|\; - dc$$

Solution.

The quadratic points must be the same in number as the single figures in the coefficient of the linear term.
The first figure must be the next greatest after the first figure of the coefficient, if the rest are consistent. If not, the same.

					c		
	b		c				
	2		4		2		
	Ȯ	0	4̇	8	4̇		
$-d$	$-$	2	4	Ȯ	·	·	
bb	4̇	·	·				
$-bd$	$-$	4	8	0			
$bb - bd$	$-$		8	0			
	·		8	4̇	8		$c = 4$
$-d$	$\dot{-}$		2	4̇	Ȯ		
$2b$			4				
$2b - d$			I	6	0		divisor
$-dc$			9	6	0		
$2bc$		I	6				
cc		I	6				
$2bc + cc$		I	7	6			
			4̇	8	4̇		$c = 2$
$-d$		$-$	2̇	4̇	Ȯ		
$2b$			4̇	8			
$2b - d$			2	4	0		divisor
$-dc$		$-$	4	8	0		
$2bc$			9̇	6	·		
cc					4		
$2bc + cc$			9	6	4		
$2bc + cc - dc$			4	8	4		
			0	0	0		

b.3)   *On solving equations in numbers*

Problem 10)[4] $aa - da = xz$

$$80 - 60 = 1600$$
$$80 \quad 80$$

Viète's third example[5]
artificium parabolae

$$aa - 60a = 1600$$

epanorthicum[6]

Canonical form:

$$b + c - d \quad = \quad bb \,\big|+ 2bc + cc$$
$$b + c \quad b + c \quad - bd \,\big|- dc$$

Solution.

The quadratic points must be the same in number as the single figures in the coefficient of the linear term.

The square of 6 is 36

$$+16$$

52   of which the nearest figure after the square root is 8, to be put for the first figure if the rest are consistent.

		b		c	
		8		0	
	1	6̇	0	0̇	
$-d$	$-$	6	0̇	.	
$-db$	-4	8̇	0̇		
bb	6	4			
$bb - db$	1	6	0		
	0	0	0	$c = 0$	

In the affirmative equation

$$aa + da = xz$$

$$8 + 8 = 128$$
$$8 \quad 8$$

Viète's fourth example[7]

$$aa + 8a = 128$$

epanorthicum

Solution.

The square of 8 is 64

$$128$$

difference $\qquad$ 64 $\qquad$ of which the square root is 8, to be put for the first figure.

$$+ d \qquad \overset{.}{1} \quad 2 \quad \overset{..}{8} \\ \qquad \qquad \quad \overset{.}{8}$$

HMC 241/1 f. 9

b.4) $\qquad$ *On solving equations in numbers*

Problem 11) $\qquad$ $aaa - dda = xxz$

$$24 - 10 = 13584$$
$$24 \quad 1$$
$$24 \quad 24$$

$aaa - 10a = 13584$ $\qquad$ Viète's first example[8]

Canonical form: $\qquad$
$$\begin{array}{l} b + c - d \\ b + c \quad\ \ d \\ b + c \quad b + c \end{array} \quad = \quad \left. \begin{array}{l} bbb \\ -\ ddb \\ \end{array} \right| \begin{array}{l} +\ 3bbc + 3bcc + ccc \\ -\ ddc \\ \end{array}$$

Solution.

			b			c	
			2			4	
		I	$\dot{3}$	5	8	$\dot{4}$	
$-dd$			$-$	I	$\dot{0}$		
bbb			8				
$-ddb$				2	0		
$bbb - ddb$			7	8	0		
			$\dot{5}$	7	8	$\dot{4}$	$c = 4$
$-dd$				$-$	i	$\dot{0}$	
$3bb$			i	2			
$3b$					6		
$3bb + 3b$			I	2	6		
$3bb + 3b - dd$			I	2	5	0	divisor
B	$-ddc$			$-$	4	0	
	$3bbc$		$\dot{4}$	8			
	$3bcc$			9	6		
	ccc			6	$\dot{4}$		
$3bbc + 3bcc + ccc$			5	8	2	4	
B			$\dot{5}$	7	8	4	
			0	0	0	0	

HMC 241/1 f. 8

b.5)   *On solving equations in numbers*

Problem 11) $aaa - dda = xxz$

$343 - 116620 = 352947$

343   I

343   343   Viète's second example[9]

$aaa - 116620a = 352947$ acephalic cube[10]

Canonical form:
$$b + c - d \quad = \quad bbb \mid + 3bbc + 3bcc + ccc$$
$$b + c \quad d \qquad - ddb \mid - ddc$$
$$b + c \quad b + c$$

Solution.

The cubic points must be the same in number as the quadratic points in the coefficient of the square term. $\sqrt{11}$ is 4, to be put for the first figure if the rest agree; if not it will be 3.

	Term	±	D1	D2	D3	D4	D5	D6	D7	Note
				b			*c*			
				3			4		3	
			Ȯ	3	5	2̇	9	4	7̇	
	− dd	−		I	I	6	6	2	Ȯ	
	− ddb	−	3	4̇	9	8̇	6	Ȯ		
	bbb		2	7						
	− ddb + bbb	−		7	9	8	6	0		
				8	3	3	8̇	9	4	c = 4
B	− dd	−		I	I	6	6	2̇	Ȯ	
	3bb			2	7					
	3b					9				
	3bb + 3b			2	7	9				
B divisor				I	6	2	3	8	0	
C	− ddc	−		4̇	6	6	4	8	0	
	3bbc			I	Ȯ	8				
	3bcc					I	4	4		
	ccc							6	4	
	3bbc + 3bcc + ccc			I	2	3	0	4		
C divisor				7	6	3	9	2	0	
			·	6	9	9̇	7	4	7̇	c = 3
B	− dd	−		I	I	6	6̇	2̇	Ȯ	
	3bb				3	4	6̇	8		
	3b						I	0	2	
	3bb + 3b				3	4	7	8	2	
B divisor				2	3	I	2	0	0	
C	− ddc	−		3	4	9	8	6	0	
	3bbc			I	0	4	Ȯ	4		
	3bcc						9	I	8	
	ccc							2	7	
	3bbc + 3bcc + ccc		I	0	4	9̇	6	0	7	
C divisor				6	9	9	7	4	7	
				0	0	Ȯ	0	0	Ȯ	

b.6)   *On solving equations in numbers*

Problem 11) $aaa - dda = xxz$

$$90 - 6400 = 153000$$

90	1
90	90

$$aaa - 6400a = 153000$$

Viète's third example[11]
artificium parabolae
epanorthicum

Canonical form:

$$
\begin{array}{llll}
b + c - d & = & bbb & + 3bbc + 3bcc + ccc \\
b + c \quad d & & - ddb & - ddc \\
b + c \quad b + c & &
\end{array}
$$

Solution.

The cubic points must be the same in number as the quadratic points in the coefficient of the square term.

$\sqrt{64}$ is 8 whose cube is 5 1 2

$$
\begin{array}{r}
+ 1\ 5\ 3 \\
\hline
6\ 6\ 5
\end{array}
$$

of which the nearest figure after the cube root is 9, to be put for the first figure if the rest are consistent.

			9			0
	1	5	3̇	0	0	0̇
$- dd$		−	6	4	0	0̇
$- ddb$	−	5	7	6̇	0	0̇
bbb		7	2	9		
$bbb - ddb$		1	5	3	0	0
		0	0	0̇	0	0

$c = 0$

In the affirmative equation
$$aaa + dda = xxz$$

$$8 + 64 = 1024$$

8	1
8	8

Viète's fourth example[12]

$$aaa + 64a = 1024 \qquad \qquad \textit{epanorthicum}$$

Solution.

$\sqrt{64}$ is 8 whose cube is 5 1 2

$$\begin{array}{r} 1\,0\,2\,4 \\ \hline 5\,1\,2 \end{array}$$

difference 5 1 2 of which the cube root, 8, is to be put for the first figure.

				8
	$\dot{1}$	0	2	$\dot{4}$
dd			$\dot{6}$	$\dot{4}$
ddb		5	1	2
bbb		5	1	2
bbb + *ddb*	1	0	2	4

b.7)   *On solving equations in numbers*

Problem 12) $aaa - daa = xxz$

$$27 - 7 = 14580$$
27 27
27 27

$aaa - 7aa = 14580$ Viète's first example[13]

Canonical form: $b + c - d \qquad = \ bbb \, | + 3bbc + 3bcc + ccc$
                     $b + c \quad b + c \qquad - \ bbd \, | - 2bdc - dcc$
                     $b + c \quad b + c$

Solution.

			b			c	
			2			7	
		I	$\dot{4}$	5	8	$\dot{0}$	
$-d$			$-$	$\dot{7}$			
$-bbd$			$-$	2	8		
bbb			8				
$bbb - bbd$			$\dot{5}$	2			
			$\dot{9}$	3	8	0	$c = 7$
B	$-d$		$\cdot$		$-$	$\dot{7}$	
	$-2bd$		$-$	2	8		
	$-d - 2bd$		$-$	2	8	7	
	$3bb$	I	2				
	$3b$			6			
	$3bb + 3b$	I	2	6			
B divisor			9	7	3		
	$-dcc$		$-$	$\dot{3}$	4	$\dot{3}$	
	$-2bdc$		$-$	I	9	6	
	$-dcc - 2bdc$		$-$	$\dot{2}$	3	0	$\dot{3}$
	$3bbc$		8	4			
	$3bcc$		2	9	4		
	ccc			3	4	3	
$3bbc + 3bcc + ccc$		I	$\dot{1}$	6	8	3	
C divisor			$\dot{9}$	3	8	$\dot{0}$	
			0	$\dot{0}$	0	$\dot{0}$	

b.8)	*On solving equations in numbers*
Problem 12)	$aaa - daa = xxz$

$$12 - 10 = 288$$
$$12 \quad 12$$
$$12 \quad 12$$

Viète's second example[14]

$$aaa - 10aa = 288$$

acephalic cube

Canonical form:

$$
\begin{array}{llll}
b + c - d & = & bbb & + 3bbc + 3bcc + ccc \\
b + c \quad b + c & & - bbd & - 2bdc - dcc \\
b + c \quad b + c & & &
\end{array}
$$

Solution.

The cubic points must be the same in number as the single figures in the coefficient of the square term.

		b			c
		I			2
		Ȯ	2	8	8̇
− d	−	I	Ȯ		
− bbd	−	I	O		
bbb		I			
bbb − bbd	±	O	O		
			2	8	8̇ $c = 2$
B − d	÷		I	Ȯ	
− 2bd	−		2	O	
− d − 2bd	−		2	I	O
3bb			3		
3b				3	
3bb + 3b			3	3	
B divisor			I	2	O
− dcc	÷			4	Ȯ
− 2bdc	−			4	O
− dcc − 2bdc	÷		4	4	Ȯ
3bbc			6		
3bcc			I	2	
ccc					8
3bbc + 3bcc + ccc			7	2	8
C divisor			2	8	8̇
			Ȯ	O	Ȯ

b.9)   *On solving equations in numbers*

Problem 12) $aaa - daa = xxz$

$$12 - 7 = 720$$

12   12

12   12   Viète's third example[15]
                                 epanorthicum artificium parabolae
$aaa - 7aa = 720$ *and acephalic cube*

Canonical form: $b + c - d$ $=$ bbb | $+ 3bbc + 3bcc + ccc$

$b + c$ $b + c$ $- bbd$ | $- 2bdc - dcc$

$b + c$ $b + c$

Solution.

$$7 \quad 2 \quad \dot{0}$$
$$-d \qquad -\dot{7}$$

By *epanorthicum*:

the cube of 7   3 4 3

$+ 7 2 0$

1 0 6 3   of which the cube root, to be put for the first
and only figure, consists of two digits, that is 10. Since 720, given, has
only one cubic point, this is absurd.

This is therefore an acephalic cube and the solution is to be carried out as in the previous case.

	b			c	
	1			2	
	$\dot{0}$	7	2	$\dot{0}$	
$-d$ —	$-$ 0	$\dot{7}$			
$-bbd$	$-$ 7				
bbb	1				
$bbb - bbd$	3				
		. 4	2	$\dot{0}$	$c = 2$
B $\quad -d$		.	$-$	$\dot{7}$	
$-2bd$		$-$ 1	4		
$-d - 2bd$		$\dot{-}$ 1	4	$\dot{7}$	
$3bb$		3			
$3b$			3		
$3bb + 3b$		3	3		
B divisor		. 1	8	$\dot{3}$	
$-dcc$		$\dot{-}$	2	$\dot{8}$	
$-2bdc$		$-$	2	8	
$-dcc - 2bdc$		$-$ 3	0	8	
$3bbc$		6			
$3bcc$		1	2		
ccc				8	
$3bbc + 3bcc + ccc$		7	2	8	
C divisor		. 4	2	$\dot{0}$	
		$\dot{0}$	0	$\dot{0}$	

In the affirmative equation:

$$aaa + 8aa = 1024 \qquad \text{Viète's fourth example[16]}$$
$$\text{\emph{epanorthicum}}$$

$$8 + 8 = 1024$$
$$8 \quad 8$$
$$8 \quad 8$$

By *epanorthicum*:

$$1\,0\,2\,4$$
$$-5\,1\,2$$
$$\overline{5\,1\,2}$$

of which the cube root, 8, is to be put for the first figure.

After devolution

$$\begin{array}{r} 8 \\ \dot{1}\ 0\ 2\ \dot{4} \\ +d \qquad +\ \dot{8} \\ \hline \end{array}$$

HMC 241/1 f. 3

b.10)   *On solving equations in numbers*

Problem 13)[17] $aaaa - daaa + fffa = xxxz$

$$32 - 68 + 202752 = 5308416$$
$$32 \quad 32 \qquad\quad 1$$
$$32 \quad 32 \qquad\quad 1$$
$$32 \quad 32 \qquad\quad 32$$

$$aaaa - 68aaa + 202752a = 5308416$$

Canonical form:
$$\begin{array}{l|l} bbbb & +\ 4bbbc + 6bbcc + 4bccc + cccc \\ -\ bbbd & -\ 3bbdc - 3bdcc -\ dccc \\ +\ bfff & +\ fffc \end{array}$$

Solution.

	1	2	3 (b)	4	5	6 (c)	7	
			3			2		
	5	3	0̇	8	4	i̇	6	
fff	2	0	2	7	5	2̇		
− d			− 6	5̇				
B bfff	6	0	8	2	5	6		
bbbb		8	1					
bfff + bbbb	6	8	9	2	5	6		
−bbbd	− 1	8	3	6				
B divisor	5	0	5	6	5	6		
		2	5̇	1	8	5	6	c = 2
C fff		2	0̇	2	7	5	2̇	
4bbb		1	0	8				
6bb				5	4			
4b					1	2		
		3	1	6	2	7	2	
− d						− 6	8	
− 3bd				− 6	1	2		
− 3bbd		− 1	8	3	6			
		− 1	8	9	7	8	8	
C divisor		1	2	6̇	4	8	4̇	
D fffc		4	0	5	5	0	4	
4bbbc		2	1	6				
6bbcc			2	1	6			
4bccc					9	6		
cccc						1	6	
		6	4	4	0	8	0	
− dccc					− 5	4	4	
− 3bdcc				− 2	4	4	8	
− 3bbdc		− 3	6	7	2			
		− 3	9	2	2	2	4	
D divisor		2	5	1	8	5	6	
		0	0	0	0	0	0	

*b.*12)   *On solving equations in numbers*

Problem 14)[18] $aaaa + daaa - fffa = xxxz$

$$32 - 10 - 200 = 1369856$$

32	32	I
32	32	I
32	32	32

$aaaa - 10aaa - 200a = 1369856$

Canonical form:

$$
\begin{array}{l|l}
bbbb & + 4bbbc + 6bbcc + 4bccc + cccc \\
+ bbbd & + 3bbdc + 3bdcc + dccc \\
- bfff & - fffc
\end{array}
$$

Solution.

			b			c		
			3			2		
	I	3	6̇	9	8	5	6̇	
d			I	0̇				
− fff				−	2	0	0̇	
B bbbd		2	7	0				
bbbb		8	I					
bbbd + bbbb	I	0	8	0				
− bbbf			−	6	0	0		
B divisor	I	0	7̇	4	0	0		
		2	9	5	8	5	6̇	c = 2
C d						i̇	0̇	
3bd					9	0		
3bbd			2	7	0			
4bbb		I	0	8				
6bb				5	4			
4b					I	2		
		I	4	I	4	3	0	
− fff				−	2	0	0	
C divisor		I	4	I	2	3̇	0̇	
D dccc						8	0	
3bdcc				3	6	0		
3bbdc			5	4	0			
4bbbc		2	I	6				
6bbcc			2	I	6			
4bccc					9	6		
cccc						I	6	
		2	9	6	2	5	6	
− fffc				−	4	0̇	0̇	
D divisor		2	9̇	5	8	5	6̇	
		0	0	0	0	0	0	

*b.*12)　　　　　*On solving equations in numbers*

Problem 15)[19]　　$aaaaa - ddaaa + fffffa = xxxxz$

$$24 - 5 + 500 = 7905504$$

24	I	I
24	24	I
24	24	I
24	24	24

$aaaaa - 5aaa + 500a = 7905504$

Canonical form:　　$bbbbb$ | $+ 5bbbbc + 10bbbcc + 10bbccc + 5bcccc + ccccc$
　　　　　　　　　　$- bbbdd$ | $- 3bbddc - 3bddcc - ddccc$
　　　　　　　　　　$+ bffff$ | $+ ffffc$

Solution.

	c1	c2	c3	c4	c5	c6	c7	
		b					c	
		2					4	
	7	9̇	O	5	5	O	4̇	
ffff					5	O	Ȯ	
− dd					− 5̇			
B bffff				I	O	O	O	
bbbbb		3	2					
bffff + bbbbb		3	2	I	O	O	O	
− bbdd				− 4	O			
B divisor		3	i̇	7	O	O	O	
	4	7̇	3	5	5	O	4	c = 4
C ffff					5	Ȯ	Ȯ	
5bbbb			8	O				
10bbb			8	O				
10bb			4	O				
5b				I	O			
		8	8	4	6	O	O	
− dd							− 5	
− 3bdd					− 3	O		
− 3bbdd				− 6	O			
				− 6	3	O	5	
C divisor		8̇	7	8	2	9	5̇	
D ffffc				2	O	O	O	
5bbbbc		3	2	O				
10bbbcc	I	2	8	O				
10bbccc		2	5	6	O			
5bcccc			2	5	6	O		
ccccc				I	O	2	4	
	4	7̇	6	4	6	2	4̇	
− ddccc					− 3	2	O	
− 3bddcc				− 4	8	O		
− 3bbddc			− 2	4	O			
			− 2	9	I	2	O	
D divisor	4	7̇	3	5	5	O	4̇	
	O	O	O	O	O	O	O	

Fig. 10   Sheet [c].8) of the *Treatise on equations*, Add MS 6782, f. 410.

On solving equations in numbers

Add MS 6782 f. 417

$c.1$)	*On solving equations in numbers*	
Problem 16)[1]	$xz = da - aa$	for two values of a

Canonical form for	$bc =$	ba		$a = b$
unequal roots:[2]		$+ ca - aa$		$a = c$

For, if $a = b$ we will have: $bc =$ bb

$+ bc - bb$ and it is so;

if $a = c$ we will have: $bc =$ bc

$+ cc - cc$ and it is.

Therefore $a = b$ or $a = c$

Canonical form for	$bb =$	ba		$a = b$
equal roots:		$+ ba - aa$		$a = b$
or:	$bb = 2ba - aa$			

If $b + c = d$

we will have: $bc = xz$

Let b be the smaller root, c the larger.

$$2b < b + c < 2c$$
$$b < \frac{b + c}{2} < c$$

87

therefore $\ b < d < c$
$$\frac{}{2}$$

$bb < bc < cc$

$b < \sqrt{bc} < c$

therefore $\quad b < \sqrt{xz} < c$

$d: 2xz = \text{I}: \dfrac{2xz}{d}$

I say that: $\ b < \dfrac{2xz}{d} < c$

$bd < 2xz < cd$

$bb + bc < 2bc < bc + cc$

It is so. Therefore the proposition is true.

$xz = da - aa$

therefore: $da > aa$
$$d > a$$

This is true whatever the roots.

Therefore the root [never] has more figures than are in d.

If one root is known, the other will be found.

Suppose b is known, c may be sought.
$$d = b + c$$
therefore: $d - b = c$

or: $xz = bc$

therefore: $\dfrac{xz}{b} = c$

Suppose c is known, b may be sought.
$$d - c = b$$

and $\quad \dfrac{xz}{c} = b$

For example, to aid the solution.

In numbers let $b = 27, c = 343$.

Then
$$\begin{array}{r|l} 27 & = 27a \\ \hline 343 & +343a - aa \end{array}$$

That is: $9261 = 370a - aa$

From what has gone before, the limits on the roots are:

$$\left.\begin{array}{l} b < \dfrac{370}{2} = 185 < c \\[2mm] b < \sqrt{9261} = 96 < c \\[2mm] b < \dfrac{\begin{array}{|r} 9261 \\ \hline \ \ 2 \end{array}}{370} = 50\tfrac{2}{37} < c \end{array}\right\} \quad \text{and} \quad c < 370$$

Add MS 6782 f. 416

c.2)   *On solving equations in numbers*

Problem 16) $xz = da - aa$ for two values of a

$$\begin{array}{rl} bc = & ba \\ & + ca - aa \\ 9261 = & 370a - aa \end{array} \qquad\qquad \begin{array}{l} a = b \\ a = c \\ a = 27 \\ a = 343 \end{array}$$

Canonical form for the solution:
$$\dfrac{d}{b+c} \left|\begin{array}{l} -b+c \\ b+c \end{array}\right. = \dfrac{db}{-bb} \left|\begin{array}{l} + dc \\ -2bc - cc \end{array}\right.$$

Solution for the smaller root.

```
                          b        c
                          2        7
                  9       2̇   6    i
      d                   3    7   ȯ   ·
      db                  7    4   0
    − bb                    − 4
   db − bb                7    0   0

                          2    2̇  6    i        c = 7
          d                    3    7   ȯ
        − 2b                     − 4
Divisor   d − 2b               3    3   0
B              dc         2    5̇  9    ȯ
             − 2bc             − 2    8
             − cc                  − 4    9
                               − 3    2    9
B                         2    2̇  6    i
                          0    0    ȯ    ȯ
```

Therefore the smaller root is 27.

Therefore the larger root is $370 - 27 = 343$

$$\text{or: } \frac{9261}{27} = 343$$

Add MS 6782 f. 416 col. 2

Solution for the larger root.

			b	c				
			3	4	3			
			Ȯ	9	2̇	6	i̇	
	d		3	7	Ȯ	.	.	
	db	I	I	I	O			
	$-bb$		$-$	9				
	$db - bb$		2	I	O			
			$-$ i̇	I	7̇	3	9̇	$c = 4$
	d		3	7̇	Ȯ	.		
	$-2b$		$-$	6				
Divisor	$-2b + d$		$-$	2	3	O		
$-B$	dc		I	4	8	O		
	$-2bc$		$-$	2	4			
	$-cc$		$-$	I	6			
			$-$	2	5	6		
	$-B$		$-$	I	O	8	O	
				$-$ 9̇	3	9̇		$c = 3$
	d			3̇	7̇	Ȯ		
	$-2b$			$-$	6	8		
Divisor	$-2b + d$			$-$	3	I	O	
$-C$	dc			I	I	I	O	
	$-2bc$			$-$	2	O	4	
	$-cc$					$-$	9	
				$-$	2	O	4	9
	$-C$			$-$	9	3	9	
				O	O	O		

Therefore the larger root is 343.

Therefore the smaller root is $370 - 343 = 27$

$$\text{or: } \frac{9261}{343} = 27$$

c.3)   *On solving equations in numbers*

Problem 17)[3] $xxz = dda - aaa$ for two values of *a*

Canonical form for $bbc = \quad bba$

unequal roots: $+\, bcc \quad +\, bca$ $a = b$

                       $+\, cca - aaa$ $a = c$

For, if $a = b$ we will have: $bbc = \quad bbb$

                     $+\, bcc \quad +\, bbc$

                           $+\, bcc - bbb$ and it is so;

     if $a = c$ we will have: $bbc = \quad bbc$

                     $+\, bcc \quad +\, bcc$

                           $+\, ccc - ccc$ and it is.

Therefore $a = b$ or $a = c$.

Canonical form for $a = b$

equal roots: $2bbb = 3bba - aaa$ $a = b$

If $bb + bc + cc = dd$

we will have: $bbc + bcc = xxz$.

bb, bc, cc are in continued proportion.

Let *b* be the smaller root, *c* the larger.

$$3bb < bb + bc + cc < 3cc$$

$$bb < \frac{bb + bc + cc}{3} < cc$$

$$bb < \frac{dd}{3} < cc$$

Therefore: $b < \sqrt{\dfrac{dd}{3}} < c$

$2bbb < bbc + bcc < 2ccc$

$bbb < \dfrac{bbc + bcc}{2} < ccc$

$bbb < \dfrac{xxz}{2} < ccc$

Therefore: $b < \sqrt[3]{\dfrac{xxz}{2}} < c$

$2bb \quad < \quad bc + cc$

$\qquad\qquad bb + bc < 2cc$

$2bbb \quad < \quad bbc + bcc$

$\qquad\qquad bbc + bcc < 2ccc$

$2bbb \quad < \quad bbc + bcc < 2ccc$

$2bbb + 2bbc + 2bcc < 3bbc + 3bcc < 2bbc + 2bcc + 2ccc$

$b < \dfrac{3bbc + 3bcc}{2bb + 2bc + 2cc} < c$

Therefore: $b < \dfrac{3xxz}{2dd} < c$

$xxz \quad = \quad dda - aaa$

$dda \quad > \quad aaa$

$dd \quad > \quad aa$

Therefore: $\sqrt{dd} > a$.

Therefore whatever the root it never has more figures than are in $\sqrt{dd}$.

---

c.4)	*On solving equations in numbers*	
Problem 17)	$xxz = dda - aaa$	two values of a

Canonical form for	$bbc =$	bba	
unequal roots:	$+ bcc$	$+ bca$	$a = b$
		$+ cca - aaa$	$a = c$

If one root is known, the other will be found.

Suppose b is known, c may be sought.

$$dd = bb + bc + cc$$
$$dd - bb = bc + cc$$

therefore c is given.

Or: $\quad xxz \quad = bbc + bcc$

$\qquad \dfrac{xxz}{d} \quad = bc + cc$

therefore c is given.

Suppose c is known, b may be sought.

$$dd - cc = cb + bb$$

therefore b is given.

Or: $\quad \dfrac{xxz}{c} = cb + bb$

therefore b is given.

For example, to aid the solution.
In numbers let $b = 12, c = 108$.

That is: $155520 = 13104a - aaa$

The limits on the roots are:

$$b < \sqrt{\frac{13104}{3}} = \sqrt{4368} < c$$

$$b < \sqrt[3]{\frac{155520}{2}} = \sqrt[3]{77760} < c$$

$$b < \left| \begin{array}{c} 155520 \\ \hline \begin{array}{c} 3 \\ \hline 13104 \\ \hline 2 \end{array} \end{array} \right| = \frac{466560}{26208} < c$$

and $\quad c < \sqrt{13104}$

*c.*5)   *On solving equations in numbers*

Problem 17) $xxz = dda - aaa$ two values of *a*

$bbc =$	bba	
$+ \ bcc$	$+ \ bca$	$a = b$
	$+ \ cca - aaa$	$a = c$
$155520 =$	$13104a - aaa$	$a = 12$
		$a = 108$

Canonical form
for the solution:

$$\left. \begin{array}{c} dd \\ \hline b + c \end{array} \right| \begin{array}{c} - \quad b + c \\ b + c \\ b + c \end{array} \right| = \begin{array}{c} ddb \\ - \ bbb \end{array} \left| \begin{array}{c} + \ ddc \\ - \ 3bbc - 3bcc - ccc \end{array} \right.$$

Solution for the smaller root.

					b	c	
					I	2	
	I	5	5̇	5	2̇	O	
dd	I	3	I	O	4̇		
ddb	I	3	I	O	4		
− bbb		− I					
ddb − bbb	I	3	O	O	4		
		2	5̇	4	8	Ȯ	c = 2
dd		I	3	I	Ȯ	4̇	
− 3bb			− 3				
− 3b				− 3			
Divisor B		I	2̇	7	7	4̇	
C + ddc		2	6	2	O	8	
− 3bbc			− 6				
− 3bcc			− I	2			
− ccc						− 8	
				− 7	2	8	
Divisor C		2	5	4	8	Ȯ	
		O	O	O	O	O	

Therefore the smaller root is 12.

Let b be the smaller root, c the larger,
and let c be sought.

$$dd - bb = bc + cc$$

or   (A)

$$\frac{xxz}{b} = bc + cc$$

$$12960 = 12c + cc$$

therefore c = 108.

A.   Proof that $dd - bb = \dfrac{xxz}{b}$

[or that] $(bb + bc + cc) - bb = \dfrac{bbc + bcc}{b}$

that is: $bc + cc = bc + cc$.

It is so. Therefore the proposition is true.

Also $dd - cc = \dfrac{xxz}{c}$

[or] $bb + bc + cc - cc = \dfrac{bbc + bcc}{c}$

that is: $bb + bc = bb + bc$.

It is so. Therefore the proposition is true.

Add MS 6782 f. 412

c.6)   *On solving equations in numbers*

Problem 17) $xxz = dda - aaa$ for two values of a

$bbc =$	bba	
$+ \, bcc$	$+ \, bca$	$a = b$
	$+ \, cca - aaa$	$a = c$
$155520 = 13104a - aaa$		$a = 12$
		$a = 108$

Canonical form for
the solution:

$$\begin{vmatrix} d & - & b + c \\ d & & b + c \\ b + c & & b + c \end{vmatrix} = \begin{array}{l} ddb \\ - \, bbb \end{array} \Big| \begin{array}{l} + \, ddc \\ - \, 3bbc - 3bcc - ccc \end{array}$$

Solution for the larger root.

		b				c		c
		1				0		8
		ȯ	1	5	5̇	5	2	ȯ
	dd	1	3	1	0	4̇		
	ddb	1	3	1	0	4		
	− bbb	− 1						
	ddb − bbb		3	1	0	4		
B	÷	1	5	4̇	8	8	ȯ	c = 0
	dd	1	3	1	0	4̇		
	− 3bb	− 3						
	− 3b		− 3					
Divisor B		− 1	9	8	9	6		
C		− 1	5	4	8	8	ȯ	c = 8
	dd		1	3	1	0	4̇	
	− 3bb		− 3	0	0			
	− 3b			− 3	0			
Divisor C			− 2	7	1	9	6̇	
D	ddc		1	0	4	8	3	2̇
	− 3bbc		− 2	4	0	0		
	− 3bcc		− 1	9	2	0	1	
	− ccc			− 5	1	2		
			− 2	5	9	7	1	2
Divisor D			− 1	5	4	8	5	0
			0	0	0	0	0	0

Therefore the larger root is 108. Now the smaller root may be sought.
Let b be the smaller root, c the larger.

$$dd - cc = cb + bb$$

or: $$\frac{xxz = cb + bb}{c}$$

$1440 = 108b + bb$
therefore $b = 12$.

c.7)   *On solving equations in numbers*

Problem 18)[4] $\qquad$ $xxz = daa - aaa$ $\qquad$ for two values of *a*

Canonical form
for unequal roots:

$$\frac{bbcc}{b+c} = \begin{array}{l} bbaa \\ + bcaa \\ \underline{+ ccaa - aaa} \\ b+c \end{array} \qquad \begin{array}{l} a = b \\ a = c \end{array}$$

that is: $bbcc =$ $\begin{array}{l} bbaa \\ + bcaa - baaa \\ + ccaa - caaa \end{array}$ $\qquad \begin{array}{l} a = b \\ a = c \end{array}$

For, if $a = b$ we will have: $bbcc =$ $\begin{array}{l} bbbb \\ + bbbc - bbbb \\ + bbcc - bbbc \end{array}$ and it is so;

if $a = c$ we will have: $bbcc =$ $\begin{array}{l} bbcc \\ + bccc - bccc \\ + cccc - cccc \end{array}$ and it is.

Therefore $a = b$ or $a = c$.

Canonical form for equal roots: $\qquad$ $\dfrac{bbb}{2} = \dfrac{3baa}{2} - aaa$ $\qquad \begin{array}{l} a = b \\ a = b \end{array}$

If: $\dfrac{bb + bc + bcc}{b+c} = d$

we will have: $\dfrac{bbcc}{b+c} = xxz$

---

bb, *bc*, *cc*, are in continued proportion

and $\dfrac{bb}{b+c}$, $\dfrac{bc}{b+c}$, $\dfrac{cc}{b+c}$ are in continued proportion.

Let b be the smaller root, c the larger.

$bb + bc < 2cc$

$$2bb < bc + cc$$

$$3bb + 3bc < 2bb + 2bc + 2cc < 3bc + 3cc$$

$$b \quad < \quad \frac{2bb + 2bc + 2cc}{3b + 3c} \quad < c$$

therefore: $\quad b \quad < \quad \dfrac{2d}{3} \quad < \quad c$

$bb + bc < 2cc$

$$2bb < bc + cc$$

$$bbbb + bbbc < 2bbcc < bccc + cccc$$

$$bbb < \frac{2bbcc < ccc}{b + c}$$

$$b < \sqrt[3]{2xxx} < c$$

I say also that: $\quad b < \sqrt{\dfrac{3xxxz}{d}} < c$

that is: $\quad bb < \dfrac{3xxxz < cc}{dd}$

as it may be written; this is therefore:

$$bbd < 3xxxz < ccd$$

$$\frac{bbbb + bbbc + bbcc}{b + c} < \frac{3bbcc}{b + c} < \frac{bbcc + bccc + cccc}{b + c}$$

It is so. Therefore the proposition is true.

$xxz = daa - aaa$

therefore $d > a$. This is so whatever the root.

c.8)   *On solving equations in numbers*

Problem 18) $xxz = daa - aaa$ for two values of a

Canonical form for unequal roots:

$$\frac{bbcc =}{b+c} \quad \begin{array}{l} bbaa \\ + bcaa \end{array}$$

$$\begin{array}{ll} + ccaa \ - aaa & a = b \\ \overline{ b+c} & a = c \end{array}$$

If one root is known, the other can be found.
Suppose b is known, c may be sought.

$d = \dfrac{bb + bc + cc}{b+c}$

$db + dc = bb + bc + cc$

$db - bb = bc - dc + cc$

Let $d - b = f$
therefore: $fb = -fc + cc$
$cc - fc = fb$
therefore c is given.

Or: $xxz = \dfrac{bbcc}{b+c}$

$xxzb + xxzc = bbcc$

$xxzb = -xxzc + bbcc$

$\dfrac{xxzb}{bb} = \dfrac{-xxzc + cc}{bb}$

Let $\dfrac{xxz}{bb} = f$

therefore: $fb = -fc + cc$
therefore c is given.

Suppose c is known, b may be sought.
$db - bb - cb = -dc + cc$

Let $d - c = g$
therefore: $gb - bb = -gc$
therefore: $gc = -gb + bb$
therefore b is given.

Or: $xxzb + xxzc = bbcc$

$xxzc = -xxzb + ccbb$

$$\frac{xxzc}{cc} = \frac{-xxzb}{cc} + bb$$

Let $\dfrac{xxz}{cc} = g$

therefore: $gc = -gb + bb$

or: $bb - gb = gc$

therefore b is given.

For example, to aid the solution.

In numbers let $b = 30,\ c = 45$.

That is: $24300 = 57aa - aaa$

The limits on the roots are:

$$b < \frac{57}{\frac{2}{3}} = \frac{114}{3} < c$$

$$b < \sqrt[3]{\frac{24300}{2}} = \sqrt[3]{48600} < c \quad\Big\}\ \text{and}\quad c < 57$$

$$b < \sqrt{\frac{24300}{\frac{3}{57}}} = \sqrt{1278+} < c$$

c.9) *On solving equations in numbers*

Problem 18) $\qquad xxz = daa - aaa \qquad$ for two values of a

---

Canonical form for unequal roots:

$$\dfrac{bbcc}{b+c} = \begin{array}{ll} bbaa \\ + bcaa \\ + ccaa - aaa & a = b \\ \overline{ \ b+c} & a = c \end{array}$$

A method different from that above.

If one root is known, the other will be found.

$$d \;=\; \dfrac{bb + bc + cc}{b+c} \quad \text{with } bb, bc, cc \text{ in continued proportion.}$$

$$b \;=\; \dfrac{bb + bc}{b+c} \quad \text{(first and second)}$$

$$c \;=\; \dfrac{bc + cc}{b+c} \quad \text{(second and third)}$$

Let $\quad f = d - b = \dfrac{cc}{b+c} \quad$ (third)

and $\quad g = d - c = \dfrac{bb}{b+c} \quad$ (first)

Suppose b is known, c may be sought.

Let there be three continued proportionals:

$$f, \qquad e, \qquad \dfrac{ee}{f},$$

[such that] $\quad f + e = c$ and $e + \dfrac{ee}{f} = b$

$b = e + \dfrac{ee}{f}$

$fb = fe + ee$

therefore e is given.

$f + e = c$

therefore c is given.

But Viète does it this way from the same premises:[5]

Let f, $\quad c - f$, $\quad \dfrac{cc - 2cf + ff}{f}$,

be three quantities in continued proportion such that

$b = c - f + \dfrac{cc - 2cf + ff}{f}$

$bf = fc - ff + cc - 2cf + ff$

therefore: $bf = -fc + cc$

the same as on the other sheet [c.8)].

Suppose c is known, b may be sought.

Let g, $\quad e$, $\quad \dfrac{ee}{g}$,

be three quantities in continued proportion such that:

$g + e = b \quad$ and $\quad e + \dfrac{ee}{g} = c$

$c = e + \dfrac{ee}{g}$

$gc = ge + ee$

therefore e is given.

$g + e = b$

therefore e is given.

But Viète does it this way:[6]

Let g, $\quad b - g$, $\quad \dfrac{bb - 2gb + gg}{g}$,

be three quantities in continued proportion such that

$c = b - g + \dfrac{bb - 2gb + gg}{g}$

Then $gc = -gb + bb$

the same as our method on the other sheet [c.8)].

---

c.10)   *On solving equations in numbers*

Problem 18) $xxz = daa - aaa$ for two values of a

Canonical form for unequal roots:

$$\frac{bbcc}{b+c} = \begin{array}{l} bbaa \\ + bcaa \\ + ccaa - aaa \\ \hline b+c \end{array} \qquad \begin{array}{l} a = b \\ a = c \end{array}$$

$$24300 = 57aa - aaa \qquad \begin{array}{l} a = 30 \\ a = 45 \end{array}$$

Canonical form for the solution:

$$\left. \begin{array}{c} d \\ b+c \\ b+c \end{array} \right| \left. \begin{array}{c} -b+c \\ b+c \\ b+c \end{array} \right| = \left. \begin{array}{c} ddb \\ -bbb \end{array} \right| \begin{array}{l} + 2dbc + dcc \\ - 3bbc - 3bcc - ccc \end{array}$$

Solution for the smaller root.

$$
\begin{array}{rrrrr}
 & b & & c & \\
 & 3 & & 0 & \\
2 & \dot{4} & 3 & 0 & \dot{0} \\
\end{array}
$$

d	5	$\dot{7}$	
ddb	5	1	3
$-bbb$	2	7	
$ddb - bbb$	2	4	3

$$0 \quad \dot{0} \quad 0 \quad 0 \quad \dot{0} \qquad c = 0$$

Therefore the smaller root is 30.
Now the larger root may be sought.
Let b be the smaller root, c the larger.

$$cc - fc = fb \quad \text{where} \quad f = d - b$$

so $cc - fc = \dfrac{xxz}{b}$

$$cc - 27c = 810$$

therefore $c = 45$.

Solution for the larger root.

Label	mult	sign		b			c	
				4			5	
			2	$\dot{4}$	3	0	$\dot{0}$	
d				5	$\dot{7}$			.
ddb	9			1	2			
$-bbb$	6	−		4				
$ddb - bbb$	2			7	2			
B $\quad -$				$\dot{2}$	9	0	$\dot{0}$	$c = 5$
					5	$\dot{7}$		
d				4	5	6		
$-2db$		−		$\dot{4}$	8	1		
$-3bb$				−	1	2		
$-3b$								
Divisor B $\quad -$				3	0	3		
C $\quad dcc$			1	$\dot{4}$	2	$\dot{5}$		
$2dbc$	2		2	8	0			
$-3bbc$	2	−	4	0				
$-3bcc$		−	3	0	0			
$-ccc$		−		1	2	5		
Divisor C			$\dot{2}$	9	0	0		
			0	0	0	0		

Therefore the larger root is 45.

If the smaller root is sought

$bb - gb = gc$ where $d - c = g$

$bb - gb = \dfrac{xxz}{c}$

$bb - 12b = 540$

therefore: $b = 30$.

c.11)   *On solving equations in numbers*

Problem 19)[7] $xxz = ddda - aaaa$ for two values of *a*

Canonical form for $bbbc =$ $bbba$ $a = b$
unequal roots: $+\ bbcc$ $+\ bbca$ $a = b$
                      $+\ bccc$ $+\ bcca$
                            $+\ ccca - aaaa$

For, if $a = b$ we will have: $bbbc =$ $bbbb$
                          $+\ bbcc$ $+\ bbbc$
                          $+\ bccc$ $+\ bbcc$
                               $+\ bccc - bbbb$ and it is so;
if $a = c$ we will have: $bbbc =$ $bbbc$
                          $+\ bbcc$ $+\ bbcc$
                          $+\ bccc$ $+\ bccc$
                               $+\ cccc - cccc$ and it is.

Therefore: $a = b$ or $a = c$.

Canonical form for equal roots: $3bbbb = 4bbba - aaaa$ $a = b$
                                                        $a = b$

If $bbb + bbc + bcc + ccc = ddd$
we will have: $bbbc + bbcc + bccc = xxxz$.

---

bbb, bbc, bcc, ccc are in continued proportion
and so are $bbbc, bbcc, bccc$.

Let b be the smaller root, c the larger.

$4bbb < bbb + bbc + bcc + ccc < 4ccc$

$$bbb < \frac{bbb + bbc + bcc + ccc}{4} < ccc$$

therefore: $b < \sqrt[3]{\dfrac{ddd}{4}} < c$

$3bbbb < bbbc + bbcc + bccc < 3cccc$

$$bbbb < \frac{bbbc + bbcc + bccc}{3} < cccc$$

therefore: $b < \sqrt[4]{\dfrac{xxxz}{3}} < c$

I say also that:

$$b < \frac{4xxxz}{3ddd} < c$$

that is: $b < \dfrac{4bbbc + 4bbcc + 4bccc}{3bbb + 3bbc + 3bcc + 3ccc} < c$

$3bbbb + 3bbbc + 3bbcc + 3bccc$ (b times the denominator)
$< 4bbbc + 4bbcc + 4bccc$
$= 3bbbc + 3bbcc + 3bccc$
$\quad + bbbc + bbcc + bccc$
$< 3bbbc + 3bbcc + 3bccc + 3cccc$ (c times the denominator)

It is so. Therefore the proposition is true.

$xxxz$	$=$	$ddda - aaaa$
$ddda$	$>$	$aaaa$
ddd	$>$	aaa

Therefore: $\sqrt[3]{ddd} > a$.

This is so whatever the root.

---

*c.*12)   *On solving equations in numbers*

Problem 19) $xxxz = ddda - aaaa$ for two values of *a*

Canonical form for unequal roots:

$$\begin{aligned} bbbc &= && bbba && a = b \\ &+ bbcc &&+ bbca && a = b \\ &+ bccc &&+ bcca \\ &&&+ ccca - aaaa \end{aligned}$$

If one root is known, the other will be found.
Suppose *b* is known, *c* may be sought.

$$ddd = bbb + bbc + bcc + ccc$$
$$ddd - bbb = bbc + bcc + ccc$$

therefore *c* is given.

Or: $xxxz = bbbc + bbcc + bccc$

$$\frac{xxxz}{b} = bbc + bcc + ccc$$

therefore *c* is given.

Suppose *c* is known, *b* may be sought.

$$ddd - ccc = ccb + cbb + bbb$$

therefore *b* is given.

Or: $xxxz = ccb + cbb + bbb$

$$\frac{xxxz}{c}$$

therefore *b* is given.

For example, to aid the solution.
In numbers let $b = 8$, $c = 27$.

that is: $217944 = 27755a - aaaa$.

The limits on the roots are:

$$b < \sqrt[3]{\frac{27755}{4}} = \sqrt[3]{6938} < c$$

$$b < \sqrt[4]{\frac{217944}{23}} = \sqrt[4]{72648} < c$$

$$\left.\begin{array}{c} b < 217944 \\ \hline 4 \\ \hline 27755 \\ \hline 3 \end{array} = \begin{array}{c} 871776 \\ \hline 83265 \end{array} < c\right\} \quad \text{and} \quad c < \sqrt[3]{27755}$$

Add MS 6782 f. 405

*c.*13)   *On solving equations in numbers*

Problem 19) $xxxz = ddda - aaaa$ for two values of a

$bbbc =$	$bbba$
$+ bbcc$	$+ bbca$
$+ bccc$	$+ bcca$
	$+ ccca - aaaa$

$a = b$
$a = c$

$217944 = 27755a - aaaa$

$a = 8$
$a = 27$

Canonical form
for the solution:

$$\begin{vmatrix} d & -b+c \\ d & b+c \\ d & b+c \\ b+c & b+c \end{vmatrix} = \begin{vmatrix} dddb + dddc \\ -bbbb - 4bbbc - 6bbcc - 4bccc - cccc \end{vmatrix}$$

Solution for the smaller root.

						c
		0				8
	2	i̇	7	9	4	4̇
ddd		2	7	7	5̇	5̇
dddb	2	2	2	0	4	0
− *bbbb*		−	4	0	9	6
ddbb − *bbbb*	2	i̇	7	9	4	4
		0	0	0	0	0

Therefore the smaller root is 8.
Now the larger root may be sought.
Let *b* be the smaller root, *c* the larger.
$ddd - bbb = bbc + bcc + ccc$
or: $\dfrac{xxxz}{b} = bbc + bbc + ccc$

$27243 = 64c + 8cc + ccc$
therefore $c = 27$.

Solution for the larger root.

				b					c	
				2					7	
			2	i	7	9	4	4̇		
		ddd	2	7	7	5	5̇			
		dddb	5	5	5	I	0			
		− bbbb	− I	6						
		dddb − bbbb	3	9	5	I	0			
			− I	7̇	7	I	5	6̇		c = 7
B		ddd	2	7	7	5̇	5̇			
		− 4bbb	− 3̇	2						
		− 6bb		− 2	4					
		− 4b			− 8					
Divisor	B			∸ 6	7	2	5̇			
C		dddc	I	9	4	2	8	5		
		− 4bbbc	− 2	2̇	4					
		− 6bbcc	− I	I	7	6				
		− 4bccc		− 2	7	4	4			
		− cccc			− 2	4	0	I		
Divisor	C		− I	7̇	7	I	5	6̇		
			0	0	0	0	0̇	0̇		

If the smaller root b is sought

$$ddd - ccc = ccb + cbb + bbb$$

or: $$\frac{xxxz}{c} = ccb + cbb + bbb$$

$$8072 = 7296 + 27bb + bbb$$

therefore $b = 8$.

*c.*14) *On solving equations in numbers*

Problem 20)[8] $xxz = daaa - aaaa$ for two values of a

Canonical form
for unequal roots:

$$\frac{bbbccc}{bb + bc + cc} = \begin{array}{ll} bbbaaa & \\ + bbcaaa & \\ + bccaaa & a = b \\ + cccaaa \quad - aaaa & a = c \\ \hline bb + bc + cc \end{array}$$

that is: $bbbccc = \begin{array}{ll} bbbaaa & \\ + bbcaaa - bbaaaa & \\ + bccaaa - bcaaaa & a = b \\ + cccaaa - ccaaaa & a = c \end{array}$

For, if $a = b$ we will have : $bbbccc = \begin{array}{l} bbbbbb \\ + bbbbbc - bbbbbb \\ + bbbbcc - bbbbbc \\ + bbbccc - bbbbcc \quad \text{and it is so;} \end{array}$

if $a = c$ we will have: $bbbccc = \begin{array}{l} bbbccc \\ + bbcccc - bbccccc \\ + bccccc - bcccccc \\ + cccccc - cccccc \quad \text{and it is.} \end{array}$

Therefore $a = b$ or $a = c$.

Canonical form for equal roots: $\dfrac{bbbb}{3} = \dfrac{baaa}{3} - aaaa$ $\begin{array}{l} a = b \\ a = b \end{array}$

If $\dfrac{bbb + bbc + bcc + ccc}{bb + bc + cc} = d$

we will have: $\dfrac{bbbccc}{bb + bc + cc} = xxxz$

---

bbb, bbc, bcc, ccc are in continued proportion

and so are $\dfrac{bbb}{bb + bc + cc}$, $\dfrac{bbc}{bb + bc + cc}$, $\dfrac{bcc}{bb + bc + cc}$, $\dfrac{ccc}{bb + bc + cc}$.

Let b be the smaller root, c the larger.

I say that: $\quad b < \dfrac{3d}{4} < c$

$$b < \frac{3bbb + 3bbc + 3bcc + 3ccc}{4bb + 4bc + 4cc} < c$$

$$4bbb + 4bbc + 4bcc < 3bbb + 3bbc + 3bcc + 3ccc < 4bbc + 4bcc + 4ccc$$

$$\begin{array}{ll} 3bbb + 3bbc + 3bcc < 3bbb + 3bbc + 3bcc + 3ccc < 3bbc + 3bcc + 3ccc \\ +bbb + bbc + bcc \hspace{5cm} +bbc + bcc + ccc \end{array}$$

It is so. Therefore the proposition is true.

I say also that: $b < \sqrt[4]{3xxxz} < c$

or: $bbbb < 3xxxz < cccc$

$\quad bbbb < \dfrac{3bbbccc}{bb + bc + cc} < cccc$

$\quad\quad bbbbbb + bbbbbc + bbbbcc < 3bbbccc < bbcccc + bccccc + cccccc$

It is so. Therefore the proposition is true.

I say also that: $b < \sqrt[3]{\dfrac{4xxxz}{d}} < c$

Or: $\quad bbb < \dfrac{4xxxz}{d} < ccc$

$bbbd < 4xxxz < cccd$

$\dfrac{bbbbbb + bbbbbc + bbbbcc + bbbccc}{bb + bc + cc} < \dfrac{4bbbccc}{bb + bc + cc}$

$\hspace{5cm} < \dfrac{bbbccc + bbccccc + bccccc + cccccc}{bb + bc + cc}$

It is so. Therefore the proposition is true.

$xxxz = daaa - aaaa$

Therefore $d > a$. This is so whatever the root.

c.15) On solving equations in numbers

Problem 20) $xxxz = daaa - aaaa$ for two values of a

| Canonical form for unequal roots: | $\dfrac{bbbccc}{bb + bc + cc}$ | $=$ | $\begin{aligned} &bbbaaa \\ &+ bbcaaa \\ &+ bccaaa \\ &\dfrac{+ cccaaa}{bb + bc + cc} \end{aligned} \quad - aaaa$ | $a = b$
 $a = c$ |

If one root is known, the other will be found.

Suppose b is known, c may be sought.

$$d = \frac{bbb + bbc + bcc + ccc}{bb + bc + cc}$$
$$dbb + dbc + dcc = bbb + bbc + bcc + ccc$$
$$dbb - bbb = - dbc - dcc + bbc + bcc + ccc$$

Let $d - b = f$

$fbb = - fbc - fcc + ccc$

therefore c is given.

Or: $xxxz = \dfrac{bbbccc}{bb + bc + cc}$

$xxxzbb + xxxzbc + xxxzcc = bbbccc$

$xxxzbb = - xxxzbc - xxxzcc + bbbccc$

$\dfrac{xxxzbb}{bbb} = - \dfrac{xxxzbc}{bbb} - \dfrac{xxxzcc}{bbb} + ccc$

Let $\dfrac{xxxz}{bbb} = f$

then $fbb = - fbc - fcc + ccc$

therefore c is given.

Suppose c is known, b may be sought.

Let $d - c = g$

$gcc = - gcb - gbb + bbb$

therefore b is given.

Also $xxxz = g$
$$\overline{ccc}$$

whence by the same equation:

$gcc = - gcb - gbb + bbb$

therefore b is given.

Proof

that: $xxxz = d - b$
$$\overline{bbb}$$

[or] $xxxz = dbbb = bbbb$

$$\frac{bbbccc}{bb + bc + cc} \quad = \quad \begin{array}{l} bbbbbb - bbbb \\ + bbbbbc \\ + bbbbcc \\ + bbbccc \\ \hline bb + bc + cc \end{array}$$

$bbbccc = bbbbbb + bbbbbc + bbbbcc + bbbccc - bbbbbb - bbbbbc - bbbbcc$

It is so. Therefore the proposition is true.

For example, to aid the solution.

In numbers let $b = 38, c = 57$.

that is: $1481544 = 65aaa - aaaa$.

The limits on the roots are:

$$\left. \begin{array}{l} b \quad < \quad \dfrac{65}{\dfrac{3}{4}} \quad = 48\tfrac{1}{4} < c \\[3em] b < \sqrt[4]{\dfrac{1481544}{3}} = \sqrt[4]{4444632} < c \\[3em] b < \sqrt[3]{\dfrac{1481544}{\dfrac{4}{65}}} = \sqrt[3]{91171} < c \end{array} \right\} \text{ and } \quad c < 65$$

c.16)   *On solving equations in numbers*

Problem 20) $xxxz = daaa - aaaa$ for two values of a

Canonical form
for unequal roots:

$$\frac{bbbccc}{bb + bc + cc} = \begin{array}{l} bbbaaa \\ + bbcaaa \\ + bccaaa \\ + cccaaa \\ \hline bb + bc + cc \end{array} - aaaa \qquad \begin{array}{l} a = b \\ a = c \end{array}$$

Another method than that above, and like Viète's.
If one root is known, the other will be found.

$d = \dfrac{bbb + bbc + bcc + ccc}{bb + bc + cc}$ and

bbb, bbc, bcc, ccc are four continued proportionals.

$b = \dfrac{bbb + bbc + bcc}{bb + bc + cc}$ (1st, 2nd, 3rd)

$c = \dfrac{bbc + bcc + ccc}{bb + bc + cc}$ (2nd, 3rd, 4th)

Let $f = d - b = \dfrac{ccc}{bb + bc + cc}$ (4th)

and $g = d - c = \dfrac{bbb}{bb + bc + cc}$ (1st)

Suppose b is known, c may be sought.
Suppose there are four continued proportionals:

$$f, \quad e, \quad \frac{ee}{f}, \quad \frac{eee}{ff},$$

such that $f + e + \dfrac{ee}{f} = c$ and $e + \dfrac{ee}{f} + \dfrac{eee}{ff} = b$

$b = e + \dfrac{ee}{f} + \dfrac{eee}{ff}$

$ffb = ffe + fee + eee$

therefore e is given and hence c.

$27702 = 729e + 27ee + eee$

then $e = 18$

whence $ee = 12$
$\overline{f}$

$f = 27$

$e + ee + f = 57 = c$
$\overline{f}$

Suppose c is known, b may be sought.

Suppose $\quad g, \quad e, \quad \dfrac{ee}{g}, \quad \dfrac{eee}{gg},$

are continued proportionals such that:

$g + e + \dfrac{ee}{g} = b \quad$ and $\quad e + \dfrac{ee}{g} + \dfrac{eee}{gg} = c$

$c = e + \dfrac{ee}{g} + \dfrac{eee}{gg}$

$ggc = gge + gee + eee$

Therefore e is given and hence b.

$3648 = 64e + 8ee + eee$

then $e = 12$

whence $ee = 18$
$\overline{g}$

$g = 8$

$e + ee + g = 38 = b$
$\overline{g}$

c.17)   *On solving equations in numbers*

Problem 20) $xxz = daaa - aaaa$ for two values of a

| Canonical form for unequal roots: | $\dfrac{bbbccc}{bb + bc + cc}$ | $=$ | $\dfrac{\begin{array}{l} bbbaaa \\ + bbcaaa \\ + bccaaa \\ + cccaaa \quad - aaaa \end{array}}{bb + bc + cc}$ | $a = b$
 $a = c$ |

$$1481544 = 65aaa - aaaa$$

$a = 38$

$a = 57$

Canonical form for the solution:

$$\left. \begin{array}{c} d \\ b+c \\ b+c \\ b+c \end{array} \right| \begin{array}{c} -b+c \\ b+c \\ b+c \\ b+c \end{array} = \left. \begin{array}{c} dbbb \\ -bbbb \end{array} \right| \begin{array}{l} + 3dbbc + 3dbcc + dccc \\ - 4bbbc - 6bbcc - 4bccc - cccc \end{array}$$

Solution for the smaller root.

		c1	c2	c3	c4	c5	c6	c7	
				b				*c*	
				3				8	
		I	4	8̇	I	5	4	4̇	
	d			6	5̇				
	dbbb	I	7	5	5				
	−*bbbb*	−	8	I					
	dbbb − *bbbb*			9	4	5			
			5	3	6	5	4	4	*c* = 8
B	*d*						6	5̇	
	3*db*					5	8	5	
	3*dbb*			I	7	5	5		
	−4*bbb*		−	I	0̇	8	I		
	−6*bb*				−	5	4		
	−4*b*					−	I	2	
B divisor				6̇	7	8	9	5̇	
C	*dccc*			3	3	2	8	0	
	3*dbcc*		3	7	4	4	0		
	3*dbbc*	I	4	0	4	0			
	−4*bbbc*		−	8	6̇	4			
	−6*bbcc*		−	3	4	5	6		
	−4*bccc*			−	6	I	4	4	
	−*cccc*			−	4	0	9	6	
C divisor			5	3	6	5	4	4	
			0	0	0	0	0	0	

Therefore the smaller root is 38. The larger may be sought.
Let b be the smaller root, c the larger.

$d - b = f$

$fbb = -fbc - fcc + ccc$

$\dfrac{xxxz}{b} = -fbc - fcc + ccc$

(Viète does this another way, as above.)[9]

$38988 = -1026c - 27cc + ccc$

therefore $c = 57$.

Solution for the larger root.

			b				*c*
			5				7
	ɪ	4	8̇	ɪ	5	4	4̇ ·
d				6	5̇		
dbbb	8	ɪ	2	5			
− *bbbb* −	6	2	5				
dbbb − *bbbb*	ɪ	8	7	5			
−		3	9	3	4	5	6 *c* = 7
d					·	6	5̇
3*db*				9	7	5	
3*dbb*		4	8	7	5		
− 4*bbb* −		5	0̇	0			
− 6*bb* −		ɪ	5	0			
− 4*b*			−	2	0		
B divisor −		ɪ	7	8	8	5	
dccc		2	2	2	9	5	
3*dbcc*		4	7	7	7	5	
3*dbbc*	3	4	ɪ	2	5		
− 4*bbbc* −	3	5	0̇	0			
− 6*bbcc* −		7	3	5	0		
− 4*bccc* −		6	8	6	0		
− *cccc* −			2	4	0	ɪ	
C divisor −		3	9	3	4	5	6
		0	0	0	0	0	0

Therefore the larger root is 57. The smaller may be sought.

Let c be the larger root, b the smaller.

$$d - c = g$$
$$65 - 57 = 8$$
$$gcc = -gcb - gbb + bbb$$
$$\frac{xxxz}{c} = -gcb - gbb + bbb$$

(Viète does this another way, as above.)[10]

$$25992 = -451b - 81bb + bbb$$

therefore $b = 38$.

c.18) *On solving equations in numbers*

Problem 20) $1481544 = 65aaa - aaaa$ for two values of *a*

Absurd consequences follow in solving for the larger root if the first figure is less than or greater than 5.

			4				
	1	4	8̇	1	5	4	4̇
d			6	5̇			
dbbb	4	1	6	0			
− *bbbb*		− 2	5	6			
dbbb − *bbbb*	1	6	0	0			
		− 1	1̇	8	4	5	6̇
B *d*						6	5̇
3*db*					7	8	0
3*dbb*				3	1	2	0
− 4*bbb*				− 2	5̇	6	
− 6*bb*					− 9	6	
− 4*b*						− 1	6
B divisor				5	4	1	0

Absurd; since the residual above [−118456] has a negative sign and the divisor [5410] a positive sign. The result will also be negative. Therefore the first figure is not 4.

			6				
	1	4	8̇	1	5	4	4̇
d			6	5̇			
dbbb	1	4	0	4	0		
− *bbbb*		− 1	2	9	6		
dbbb − *bbbb*		1	0	8	0		
		4	0̇	1	5	4	4̇
B *d*						6	5̇
3*db*				1	1	7	0
3*dbb*			7	0	2	0	
− 4*bbb*				− 8	6̇	4	
− 6*bb*				− 2	1	6	
− 4*b*						− 2	4
B divisor		− 1	7	2	0	7	5

Absurd; since the residual above [401544] has a positive sign and the divisor [−172075] a negative sign. The result will also be negative. Therefore the first figure is not 6.

Therefore the first figure will be 5, by the limits found previously, just as explained by the example.

Note:

The operation of solving for the larger root is sometimes similar in part to that for the smaller root. As to be expected if both roots agree in the first figure and the number of figures, for example, if the smaller root is 23 and the larger 24. Or if the smaller is 343 and the larger 347. And thus for others. But if the roots differ in the first figure the operations will be completely different as in the preceding examples expounded in these pages.

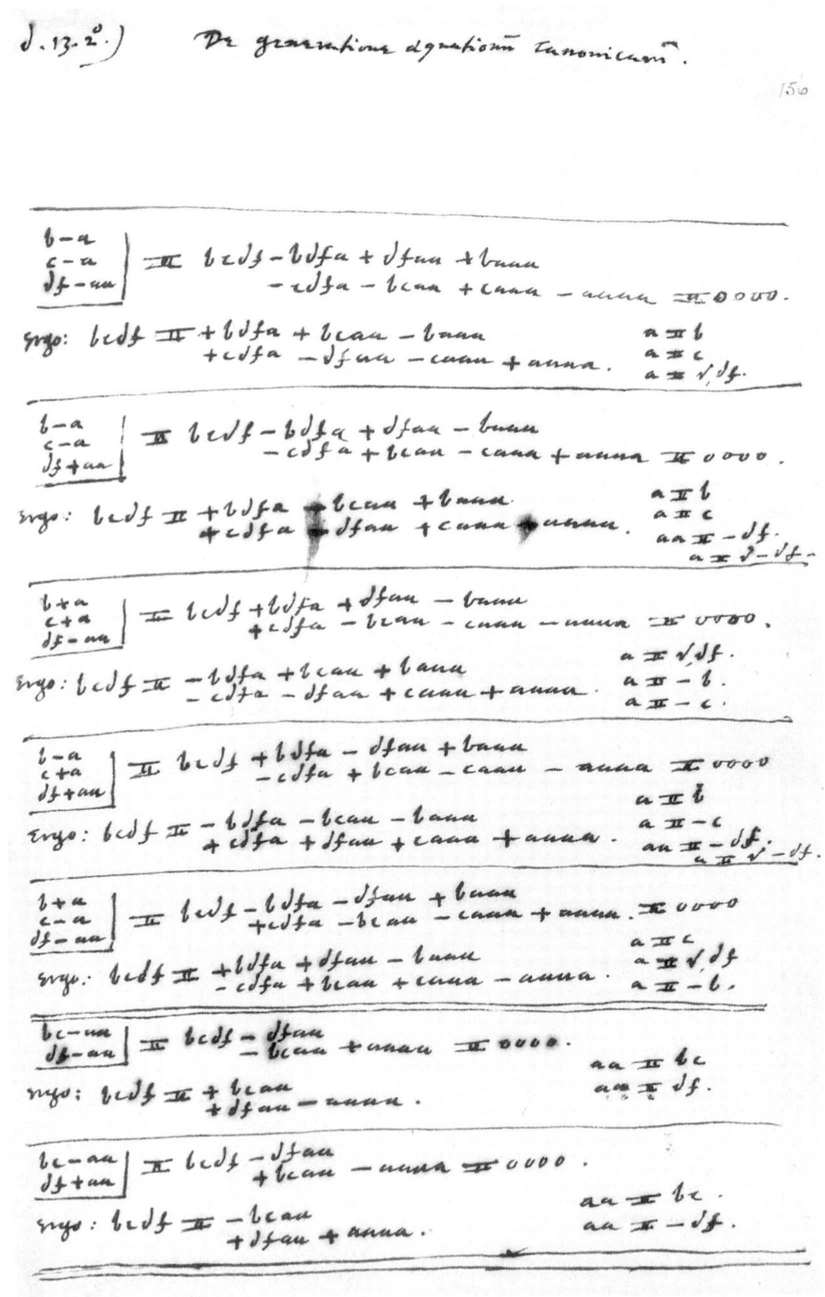

d.13.2°.) De generatione aequationum canonicarum. 156

b−a, c−a, df−aa | = bcdf − bdfa + dfaa + baaa
− cdfa − bcaa + caaa − aaaa = 0000.

Ergo: bcdf = + bdfa + bcaa − baaa
+ cdfa − dfaa − caaa + aaaa.
a = b
a = c
a = √df.

b−a, c−a, df+aa | = bcdf − bdfa + dfaa − baaa
− cdfa + bcaa − caaa + aaaa = 0000.

Ergo: bcdf = + bdfa − bcaa + baaa
+ cdfa − dfaa + caaa − aaaa.
a = b
a = c
aa = −df.
a = √−df.

b+a, c+a, df−aa | = bcdf + bdfa + dfaa − baaa
+ cdfa − bcaa − caaa − aaaa = 0000.

Ergo: bcdf = − bdfa + bcaa + baaa
− cdfa − dfaa + caaa + aaaa.
a = √df.
a = −b.
a = −c.

b−a, c+a, df+aa | = bcdf + bdfa − dfaa + baaa
− cdfa + bcaa − caaa − aaaa = 0000

Ergo: bcdf = − bdfa − bcaa − baaa
+ cdfa + dfaa + caaa + aaaa.
a = b
a = −c
aa = −df.
a = √−df.

b+a, c−a, df−aa | = bcdf − bdfa − dfaa + baaa
+ cdfa − bcaa − caaa + aaaa. = 0000

Ergo: bcdf = + bdfa + dfaa − baaa
− cdfa + bcaa + caaa − aaaa.
a = c
a = √df
a = −b.

bc−aa, df−aa | = bcdf − dfaa
− bcaa + aaaa = 0000.

Ergo: bcdf = + bcaa
+ dfaa − aaaa.
aa = bc
aa = df.

bc−aa, df+aa | = bcdf − dfaa
+ bcaa − aaaa = 0000.

Ergo: bcdf = − bcaa
+ dfaa + aaaa.
aa = bc.
aa = −df.

Fig. 11 Sheet *d.* 13.2) of the *Treatise on equations*, Add MS 6783, f. 156.

On the generation of canonical equations

Add MS 6783 f. 183

*d.*1)   *On the generation of canonical equations*

The scheme of equations
that follow:

$$a - b \quad \begin{array}{|c|} a-b \\ \hline a-c \end{array} \quad \begin{array}{|c|} a-b \\ \hline a+c \end{array} \quad \begin{array}{|c|} a-b \\ a-c \\ \hline a-d \end{array} \quad \begin{array}{|c|} a-b \\ a-c \\ \hline a+d \end{array} \quad \begin{array}{|c|} a+b \\ a+c \\ \hline a-d \end{array}$$

In the linear term $a - b$ or $b - a$ let $a = b$

therefore: $a - b = 0$

or: $b - a = 0$

therefore $a = b$ but not any d other than b.

If it were, we would have $d = b$, against the supposition, for d is supposed other than b.

Let $a = b$ in the multiplication $\begin{vmatrix} b-a \\ c-a \end{vmatrix}$ or $\begin{vmatrix} a-b \\ a-c \end{vmatrix}$
and $a = c$

therefore: $\begin{vmatrix} b-a \\ c-a \end{vmatrix} = bc - ca$
$$- ba + aa = \text{oo}$$

or: $\begin{vmatrix} a-b \\ a-c \end{vmatrix} = aa - ba$
$$- ca + bc = \text{oo}$$

therefore: $bc = \quad ba$
$$+ ca - aa$$

and we will have $a = b$ and $a = c$ if b and c are unequal, but nothing other than b and c.

If $a = b$ we will have: $bc = \quad bb$
$$+ bc - bb \qquad \text{and it is so.}$$

If $a = c$ we will have: $bc = \quad bc$
$$+ cc - cc \qquad \text{and it is.}$$

Therefore the proposition is true.

Nor will we have $a = d$ other than b and c.

If it were, we would have: $bc = \quad bd$
$$+ cd - dd$$

and: $bc - bd = cd - dd$

and: $\begin{vmatrix} c-d \\ b \end{vmatrix} = \begin{vmatrix} c-d \\ d \end{vmatrix}$

therefore $b = d$ against the proposition.

Also: $bc - cd = bd - dd$

and: $\begin{vmatrix} b-d \\ c \end{vmatrix} = \begin{vmatrix} b-d \\ d \end{vmatrix}$

therefore $c = d$ against the proposition.

Let $a = b$ in the multiplication $\begin{array}{c} b - a \\ \hline c + a \end{array}$ or $\begin{array}{c} a - b \\ \hline a + c \end{array}$

therefore: $\begin{array}{l} b - a \\ \hline c + a \end{array} = bc - ca$
$\qquad\qquad + ba - aa = 00$

or: $\begin{array}{l} a - b \\ \hline a + c \end{array} = aa - ba$
$\qquad\qquad + ca - bc = 00$

therefore: $bc = -ba$
$\qquad\qquad + ca + aa$

and we will have: $a = b$
and a is not equal to c nor anything other than b.

If $a = b$ we will have: $bc = -bb$
$\qquad\qquad\qquad\quad + bc + bb$ and it is so.

If $a = c$ we will have: $bc = -bc$
$\qquad\qquad\qquad\quad + cc + cc$
$\qquad\qquad 2bc = 2cc$

therefore $b = c$, against the proposition.
Therefore $a = b$ and not c.

Nor will we have $a = d$ other than b.

If it were, we would have: $bc = -bd$
$\qquad\qquad\qquad\qquad\quad + cd + dd$

and: $bc + bd = cd + dd$

and: $\begin{array}{l} c + d \\ \hline \quad b \end{array} = \begin{array}{l} c + d \\ \hline \quad d \end{array}$

therefore $b = d$, against the supposition, for d is supposed other than b.

If $b = c$ the first degree term is removed,
and we will have:

$$bb = aa$$

and: $a = b$

---

d.2)   *On the generation of canonical equations*

In the multiplication $\quad \left.\begin{array}{c} a - b \\ a - c \\ a - d \end{array}\right|$ or $\left.\begin{array}{c} b - a \\ c - a \\ d - a \end{array}\right|$ let $\begin{array}{c} a = b \\ a = c \\ a = d \end{array}$

therefore: $\quad \left.\begin{array}{c} a - b \\ a - c \\ a - d \end{array}\right| = aaa - baa$

$$- caa + bca$$
$$- daa + bda$$
$$+ cda - bcd = \text{ooo}$$

or: $\quad \left.\begin{array}{c} b - a \\ c - a \\ d - a \end{array}\right| = bcd - dca$

$$- dba + daa$$
$$- bca + caa$$
$$+ baa - aaa = \text{ooo}$$

therefore: $\quad bcd = bca$

$$+ bda - baa$$
$$+ cda - caa$$
$$- daa + aaa$$

and we will have $a = b$, $a = c$, $a = d$, but not f other than b, c or d.

If $\quad a = b$

then we will have: $\quad bcd = \quad bbc$
$$+ bbd - bbb$$
$$+ bcd - bbc$$
$$- bbd + bbb \qquad \text{and it is so.}$$

If $\quad a = c$

then we will have: $\quad bcd = \quad bcc$
$$+ bcd - bcc$$
$$+ ccd - ccc$$
$$- ccd + ccc \qquad \text{and it is so.}$$

If $\quad a = d$

then we will have: $\quad bcd = \quad bcd$
$$+ bdd - bdd$$
$$+ cdd - cdd$$
$$- ddd + ddd \qquad \text{and it is.}$$

Therefore the proposition is true.

a will not be not equal to f other than b, c, d.
If it were,
then we would have: $\quad bcd = \quad bcf$
$$+ bdf - bff$$
$$+ cdf - cff$$
$$- dff + fff$$

and: $\quad bcd + bff = \quad bcf$
$$+ cff \quad + bdf$$
$$+ dff \quad + cdf + fff$$

If $f = b$ or c or d then it is obvious that these equations are true.
But f is not equal to b, c or d by supposition.
Therefore f is not equal to a.

d.3)   *On the generation of canonical equations*

In the multiplication $\left.\begin{array}{c} a-b \\ a-c \\ a+d \end{array}\right|$ let $a = b$
                                                       $a = c$

therefore: $\left.\begin{array}{c} a-b \\ a-c \\ a+d \end{array}\right|$ $= aaa - baa$
$$- caa + bca$$
$$+ daa - bda$$
$$- cda + bcd = \text{ooo}$$

therefore: $bcd = - bca$
$$+ bda + baa$$
$$+ cda + caa$$
$$- daa - aaa$$

and we will have $a = b$ or $a = c$, but not $a = d$ nor f other than b and c.

If $a = b$

then we will have: $bcd = - bbc$
$$+ bbd + bbb$$
$$+ bcd + bbc$$
$$- bbd - bbb \qquad \text{and it is so.}$$

If $a = c$

then we will have: $bcd = - bcc$
$$+ bcd + bcc$$
$$+ ccd + ccc$$
$$- ccd - ccc \qquad \text{and it is.}$$

---

a is not equal to d.

If it were,

then we would have:
$$bcd = -\,bcd$$
$$+\,bdd + bdd$$
$$+\,cdd + cdd$$
$$-\,ddd - ddd$$

therefore: $2bdd + 2cdd = 2ddd + 2bcd$

therefore: $bd + cd = dd + bc$

therefore: $bd - bc = dd - cd$

that is: $\dfrac{d-c}{b} \;=\; \dfrac{d-c}{d}$

therefore $b = d$ against the proposition.

Or: $cd - bc = dd - bd$

that is: $\dfrac{d-b}{c} \;=\; \dfrac{d-b}{d}$

therefore $c = d$ also against the proposition.

Or another way:

let $bd + cd = dd + bc$

If $d = b$ or c then the equation is obviously true.

But d is not equal to b or c

therefore d is not equal to a.

And by similar reasoning to that above, f will not be equal to a, and similarly it may be proved for d.

Suppose as above that:

$$bcd = -\ bca \qquad\qquad\qquad a = b$$
$$+\ bda +\ baa \qquad\qquad a = c$$
$$+\ cda +\ caa$$
$$-\ daa -\ aaa$$

If $d = b + c$, the second degree term (in aa) is removed,

for then: $\quad bcd = bbc + bcc$
$$bd = bb + bc$$
$$cd = bc + cc$$
$$-d = -b - c$$

therefore: $\qquad bbc = -\ bca$
$$+\ bcc \quad +\ bba + baa$$
$$+\ bca + caa$$
$$+\ bca - baa$$
$$+\ cca - caa - aaa$$

therefore: $\qquad bbc = \quad bba$
$$+\ bcc \quad +\ bca$$
$$+\ cca - aaa$$

and: $\quad a = b$
$$a = c$$

demonstrated analytically both above and (c.3).

Suppose as above that:

$$bcd = - bca$$
$$+ bda + baa \qquad\qquad a = b$$
$$+ cda + caa \qquad\qquad a = c$$
$$- daa - aaa$$

If $bc = bd + cd$, the first degree term (in a) is removed,

for we will have: $\dfrac{bc}{b+c} = d$

then:
$$bcd = \frac{bbcc}{b+c} \qquad\qquad b = \frac{bb + bc}{b+c}$$
$$- bc = -\frac{bbc - bcc}{b+c} \qquad\qquad c = \frac{bc + cc}{b+c}$$
$$bd = \frac{bbc}{b+c} \qquad\qquad - d = \frac{-bc}{b+c}$$
$$cd = \frac{bcc}{b+c}$$

therefore:
$$\frac{bbcc}{b+c} = \frac{- bbca + bbaa}{} \\ - bcca + bcaa \\ + bbca + bcaa \\ \frac{+ bcca + ccaa}{b+c} - bcaa - aaa \\ \overline{b+c}$$

therefore:
$$\frac{bbcc}{b+c} = \frac{bbaa}{} \\ + bcaa \\ \frac{+ ccaa}{b+c} - aaa$$

and: $a = b$
$\ a = c$

demonstrated analytically in (c.7).

d.4)   *On the generation of canonical equations*

In the multiplication $\quad \begin{vmatrix} a+b \\ a+c \\ a-d \end{vmatrix} \quad$ let $\quad a = d$

therefore: $\quad \begin{vmatrix} a+b \\ a+c \\ a-d \end{vmatrix}$ $= aaa + baa$

$\qquad\qquad\qquad\qquad + caa + bca$

$\qquad\qquad\qquad\qquad - daa - bda$

$\qquad\qquad\qquad\qquad\qquad - cda - bcd = 000$

therefore: $\quad bcd = + bca$

$\qquad\qquad\quad - bda + baa$

$\qquad\qquad\quad - cda + caa$

$\qquad\qquad\qquad - daa + aaa$

and we will have $a = d$, but not b or c nor anything other than d.
This may be proved analytically as above.

In the above equation, if $d = b + c$ then the second degree term (in aa) is removed,

for then: $\quad bcd = bbc + bcc$

$\qquad\quad - bd = - bb - bc$

$\qquad\quad - cd = - bc - cc$

$\qquad\quad\ - d = - b - c$

therefore: $\quad\ bbc = + bca$

$\qquad\quad + bcc \quad - bba + baa$

$\qquad\qquad\qquad - bca + caa$

$\qquad\qquad\qquad - bca - baa$

$\qquad\qquad\qquad - cca - caa + aaa$

therefore: $\quad\ bbc = - bba$

$\qquad\quad + bcc \quad - bca$

$\qquad\qquad\qquad - cca + aaa$

and we will have: $\quad a = b + c$.

If $a = b + c$

then we will have:

$$bbc = -\ bbb + bbb$$
$$+\ bcc \quad -\ bbc + 3bbc$$
$$-\ bbc + 3bcc$$
$$-\ bcc + ccc$$
$$-\ bcc$$
$$-\ ccc$$

and it is so. Therefore the proposition is true.

From this equation and the other on sheet $d.3$), there arises the following rule for signs:

$$aaa - bba$$
$$-\ bca = \quad bbc$$
$$-\ cca \quad +\ bcc = \quad bba$$
$$+\ bca$$
$$+\ cca - aaa$$

$$a = b + c \qquad\qquad a = b$$
$$a = c$$

These are therefore conjugate equations.

Alternatively, in the above equation, if $d = b + c$

then: $\quad d - c = b$, so $d > c$

and then:

$$bcd = cdd - ccd$$
$$bc = cd - cc$$
$$-\ bd = -\ dd + cd$$
$$b = d - c$$

therefore:

$$cdd = +\ cda$$
$$-\ ccd \quad -\ cca + daa$$
$$-\ dda - caa$$
$$+\ cda + caa$$
$$-\ cda - daa + aaa$$

therefore:

$$cdd = -\ cca$$
$$-\ ccd \quad +\ cda$$
$$-\ dda + aaa$$

and we will have: $\quad a = d.$

This is for $d > c$.

But if $c > d$ we will have the converse equation, thus:

$$\begin{aligned}
ccd &= + cca \\
- cdd \quad &- cda \\
&+ dda - aaa
\end{aligned}$$

and we will have: $\quad a = d$
$$a = c - d$$

The equations may thus be written either way; and we have two pairs of conjugates:

$$\begin{aligned}
aaa &- cca \\
&+ cda \\
&- dda = \quad cdd \\
&\qquad\quad - ccd = + cca \\
&\qquad\qquad\quad - cda \\
&\qquad\qquad\quad + dda - aaa
\end{aligned}$$

$$a = d \qquad\qquad\qquad a = c$$
$$a = d - c$$

$$\begin{aligned}
aaa &- cca \\
&+ cda \\
&- dda = \quad ccd = + cca \\
&\qquad\quad - cdd \quad - cda \\
&\qquad\qquad\quad + dda - aaa
\end{aligned}$$

$$a = c \qquad\qquad\qquad a = d$$
$$a = c - d$$

d.5)   *On the generation of canonical equations*

In the above equation, that is:

$bcd = + bca$
$\quad\quad - bda + baa$
$\quad\quad - cda + caa$
$\quad\quad\quad - daa + aaa \quad\quad\quad\quad a = d$

if $bc = bd + cd$ then the first degree term (in a) is removed,

for then: $\quad \dfrac{bc}{b+c} = d$

then: $\quad bcd = \dfrac{bbcc}{b+c} \quad\quad\quad\quad b = \dfrac{bb + bc}{b+c}$

$\quad\quad\quad bc = \dfrac{bbc + bcc}{b+c} \quad\quad\quad c = \dfrac{bc + cc}{b+c}$

$\quad\quad - bd = -\dfrac{bbc}{b+c} \quad\quad\quad - d = -\dfrac{bc}{b+c}$

$\quad\quad - cd = -\dfrac{bcc}{b+c}$

therefore: $\quad \dfrac{bbcc}{b+c} = \begin{array}{ll} bbca & + bbaa \\ + bcca & + bcaa \\ - bbca & + bcaa \\ - bcca & + ccaa \end{array}$

$\dfrac{}{b+c} \quad - bcaa \quad + aaa$
$\quad\quad\quad\quad\quad \dfrac{}{b+c}$

therefore: $\quad \dfrac{bbcc}{b+c} = \begin{array}{l} bbaa \\ + bcaa \end{array}$

$\quad\quad\quad\quad \dfrac{+ ccaa}{b+c} \quad + aaa$

and we will have: $\quad a = \dfrac{bc}{b+c}$

Suppose: $a = \dfrac{bc}{b+c}$

then: $aa = \dfrac{bbcc}{bb + 2bc + cc}$

and: $aaa = \dfrac{bbbccc}{bbb + 3bbc + 3bcc + ccc}$

and: $\dfrac{bbcc}{b+c} = \dfrac{bbbbcc + 2bbbccc + bbcccc}{bbb + 3bbc + 3bcc + ccc}$

therefore: $bbbbcc = \quad bbbbcc$

$\qquad\qquad + 2bbbcc \quad + bbbccc$

$\qquad\qquad + bbcccc \quad + bbcccc + bbbccc$

and it is so.

Therefore it is true that $a = \dfrac{bc}{b+c}$

From this equation and the other ($d.3$) we have conjugate equations:

$aaa + bbaa$

$\quad + bcaa$

$\quad + ccaa = \dfrac{bcc}{b+c} = \dfrac{bbaa}{\begin{array}{l} + bcaa \\ + ccaa - aaa \end{array}}$
$\qquad\quad \overline{b+c} \qquad\qquad\qquad \overline{b+c}$

$a = \dfrac{bc}{b+c} \qquad\qquad\qquad a = b$

$\qquad\qquad\qquad\qquad\qquad a = c$

[for example] $aaa + 13aa = 144 = 13aa - aaa$

$\qquad\qquad\qquad a = 3 \qquad\qquad\qquad a = 4$

$\qquad\qquad\qquad\qquad\qquad\qquad\qquad a = 12$

Alternatively, if $bc = bd + cd$ then the first degree term (in a) is removed, for then: $bc - bd = cd$

then: $$b = \frac{cd}{c - d}$$

$$c = \frac{cc - cd}{c - d}$$

$$-d = \frac{-cd + dd}{c - d}$$

$$bcd = \frac{ccdd}{c - d}$$

therefore: $$\frac{ccdd}{c - d} = \begin{array}{l} + cdaa \\ + ccaa \\ - cdaa \\ - cdaa \\ + ddaa + aaa \\ \hline c - d \end{array}$$

therefore: $$\frac{ccdd}{c - d} = \begin{array}{l} + ccaa \\ - cdaa \\ + ddaa + aaa \\ \hline c - d \end{array}$$

and we will have: $a = d$.

And these equations are conjugate:

$$aaa + ccaa$$
$$- cdaa$$
$$+ \frac{ddaa}{c-d} = \frac{ccdd}{c-d} = + ccaa$$
$$- cdaa$$
$$+ \frac{ddaa}{c-d} - aaa$$

$$a = d \qquad\qquad\qquad a = c$$
$$a = \frac{cd}{c-d}$$

[for example] $aaa + 13aa = 144 = 13aa - aaa$
$$a = 3 \qquad\qquad a = 4$$
$$a = 12$$

Other conjugates may also be formed as in (d.3).

Add MS 6783 f. 178

d.6)   *On the generation of canonical equations*

Reciprocal equations

In the multiplication $\left.\begin{matrix} aa - bc \\ a - d \end{matrix}\right|$ let $aa = bc$
$$a = d$$

therefore: $aaa - daa - bca + bcd = 000$
therefore: $bcd = + bca + daa - aaa$ $\qquad a = \sqrt{bc}$
$$a = d$$

If $b = c$ the equation will be
$bbd = + bba + daa - aaa$ $\qquad\qquad a = b$
$$a = d$$

In the multiplication $\quad aa + bc\ \Big|\quad$ let $\quad a = d$
$$\ a - d\ \Big|$$

therefore: $\quad aaa - daa + bca - bcd = 000$

therefore: $\quad bcd = + bca - daa + aaa \qquad\qquad a = d$

If $b = c$ the equation will be

$bbd = + bba - daa + aaa \qquad\qquad\qquad\quad a = d$

In the multiplication $\quad aa - bc\ \Big|\quad$ let $\quad aa = bc$
$$\ a + d\ \Big|$$

therefore: $\quad aaa + daa - bca - bcd = 000$

therefore: $\quad bcd = - bca + daa + aaa \qquad\qquad a = \sqrt{bc}$

If $b = c$ the equation will be

$bbd = - bba + daa + aaa \qquad\qquad\qquad\quad a = b$

In these equations the coefficient of the linear term multiplied by the coefficient of the square term produces the given constant term.

And the lower degrees make the highest power.

In these same equations, we have reciprocal positions:

$$a \qquad aa \qquad aaa$$
$$bcd \quad bc \quad d$$

and therefore they may be called reciprocal equations.[1]

Summary

If the coefficient of the linear term multiplied by the coefficient of the square term makes the given constant term, then the root will be known without solving.

In equations where all coefficients are positive, we have it thus:[2]

$aaa + 6aa + 8a = 48 \qquad\quad a = 6/3 = 2$

$aaa + 9aa + 18a = 162 \qquad a = 9/3 = 3$

$aaa + 3aa + 2a = 6 \qquad\quad a = 3/3 = 1$

but of these and others of this kind elsewhere.

$aaa + 6aa + 9a = 54 \qquad a$ is not 2

d.7)   *On the generation of canonical equations*

$$
\begin{aligned}
a-b\,\big| &= aaaa - baaa \\
a-c\,\big| &\quad - caaa + bcaa \\
a-d\,\big| &\quad - daaa + bdaa \\
a-f\,\big| &\quad - faaa + cdaa - bcda \\
&\quad + bfaa - bcfa \\
&\quad + cfaa - bdfa \\
&\quad + dfaa - cdfa + bcdf = 0000
\end{aligned}
$$

therefore: $bcdf = +\ bcda - bcaa$

$$
\begin{aligned}
&+ bcfa - bdaa \\
&+ bdfa - cdaa + baaa && a = b \\
&+ cdfa - bfaa + caaa && a = c \\
&- cfaa + daaa && a = d \\
&- dfaa + faaa - aaaa && a = f
\end{aligned}
$$

$$
\begin{aligned}
a-b\,\big| &= aaa - baaa \\
a-c\,\big| &\quad - caaa + bcaa \\
a-d\,\big| &\quad - daaa + bdaa \\
a+f\,\big| &\quad + faaa + cdaa - bcda \\
&\quad - bfaa + bcfa \\
&\quad - cfaa + bdfa \\
&\quad - dfaa + cdfa - bcdf = 0000
\end{aligned}
$$

therefore: $bcdf = -\ bcda + bcaa$

$$
\begin{aligned}
&+ bcfa + bdaa \\
&+ bdfa + cdaa - baaa && a = b \\
&+ cdfa - bfaa - caaa && a = c \\
&- cfaa - daaa && a = d \\
&- dfaa + faaa + aaaa
\end{aligned}
$$

From this equation arise the three following:

If $b + c + d = f$ the third degree term (in aaa) is removed, and by reduction we have:

$$
\begin{aligned}
+ bbcd &= & + bbca & \\
+ bccd & & + bbda - bbaa & \\
+ bcdd & & + bcca - ccaa & \\
& & + ccda - ddaa & \qquad a = b \\
& & + bdda - bcaa & \qquad a = c \\
& & + cdda - bdaa & \qquad a = d \\
& & + 2bcda - cdaa + aaaa &
\end{aligned}
$$

If $bc + bd + cd = bf + cf + df$ the second degree term (in aa) is removed,

for then: $\quad \dfrac{bc + bd + cd}{b + c + d} = f$

whence by reduction we have:

$$
\begin{aligned}
+ bbccd &= & + bbcca & - bbaaa \\
+ bbcdd & & + bbdda & - ccaaa \\
+ \dfrac{bccdd}{b + c + d} & & + ccdda & - ddaaa \\
& & + bcdda & - bcaaa \\
& & + bccda & - bdaaa \qquad a = b \\
& & + \dfrac{bbcda}{b + c + d} & - \dfrac{cdaaa}{b + c + d} + aaaa \qquad a = c \\
& & & \qquad\qquad\qquad a = d
\end{aligned}
$$

If $bcd = bcf + bdf + cdf$ the first degree term (in a) is removed,

for then: $\quad \dfrac{bcd}{bc + bd + cd} = f$

whence by reduction we have:

$$
\begin{aligned}
\dfrac{bbccdd}{bc + bd + cd} &= & + bbccaa & - bbcaaa \\
& & + bbddaa & - bbdaaa \\
& & + ccddaa & - bccaaa \\
& & + bcddaa & - bddaaa \\
& & + bccdaa & - ccdaaa \qquad a = b \\
& & + \dfrac{bbcdaa}{bc + bd + cd} & - cddaaa \qquad a = c \\
& & & - \dfrac{2bcdaaa}{bc + bd + cd} + aaaa \qquad a = d
\end{aligned}
$$

d.7.2)^3

$$bcdf = - bcda + bcaa$$
$$+ bcfa + bdaa$$
$$+ bdfa + cdaa - baaa \qquad a = b$$
$$+ cdfa - bfaa - caaa \qquad a = c$$
$$- cfaa - daaa \qquad a = d$$
$$- dfaa + faaa - aaaa \qquad a = -f$$

To remove aa and aaa

it is required that: $\quad b + c + d = f$

[and that] $\quad bc + bd + cd = bf + cf + df$

then: $\qquad bc + bd + cd = (bb + bc + bd) + (bc + cc + cd) + (bd + cd + dd)$

therefore: $\quad bb + bc + cc = - bd - cd - dd$

and: $\qquad - bb - bc - cc = + bd + cd + dd$

therefore: $\quad 1^{\text{st}} \; d = \dfrac{-b - c}{2} + \sqrt{\dfrac{-3bb - 2bc - 3cc}{4}}$

$\qquad\qquad 2^{\text{nd}} \; d = \dfrac{-b - c}{2} - \sqrt{\dfrac{-3bb - 2bc - 3cc}{4}}$

therefore: $\quad 1^{\text{st}} \; f = \dfrac{b + c}{2} + \sqrt{\dfrac{-3bb - 2bc - 3cc}{4}}$

$\qquad\qquad 2^{\text{nd}} \; f = \dfrac{b + c}{2} - \sqrt{\dfrac{-3bb - 2bc - 3cc}{4}}$

Therefore: $\qquad d - f = - b - c$
$$- d + f = + b + c$$

and: $\qquad df = - bb - bc - cc$
$$- df = + bb + bc + cc$$

and: $\quad \left.\begin{array}{c} d - f \\ \hline b + c \end{array}\right| = - bb - 2bc - cc$

Then the coefficients of the linear term and the given constant term are reduced thus:[4]

$$- bcd + bcf = + bbc + bcc$$
$$+ bdf + cdf = - bbb - bbc - bcc$$
$$- bbc - bcc - ccc$$

$$bcdf = - bbbc - bbcc - bccc$$

Therefore the equation reduced to two terms will be:

$$- bbbc = - bbba$$
$$- bbcc \quad - bbca$$
$$- bccc \quad - bcca \qquad\qquad a = b$$
$$- ccca + aaaa \qquad\qquad a = c$$

and by change of sign it will be (as in $d.10$):

$$bbbc = bbba$$
$$bbcc \quad bbca$$
$$bccc \quad bcca \qquad\qquad a = b$$
$$ccca - aaaa \qquad\qquad a = c$$

d.8)   *On the generation of canonical equations*

$$
\left.\begin{array}{l} a+b \\ a+c \\ a+d \\ a-f \end{array}\right|
\begin{array}{l}
= aaaa + baaa \\
\quad + caaa + bcaa \\
\quad + daaa + bdaa \\
\quad - faaa + cdaa + bcda \\
\qquad - bfaa - bcfa \\
\qquad - cfaa - bdfa \\
\qquad - dfaa - cdfa - bcdf = 0000
\end{array}
$$

therefore:
$$
\begin{array}{ll}
bcdf = & + bcda + bcaa \\
& - bcfa + bdaa \\
& - bdfa + cdaa + baaa \\
& - cdfa - bfaa + caaa \\
& \qquad - cfaa + daaa \\
& \qquad - dfaa - faaa + aaaa \qquad\qquad a = f
\end{array}
$$

(It may be asked whether the following equations are conjugates of the comparable equations in *d.*7).

From this equation arise the three following:

If $b + c + d = f$ the third degree term (in *aaa*) is removed, and by reduction the equation becomes:

$$
\begin{array}{ll}
+ bbcd = & - bbca \\
+ bccd & - bbda - bbaa \\
+ bcdd & - bcca - ccaa \\
& - ccda - ddaa \\
& - bdda - bcaa \\
& - cdda - bdaa \\
& - 2bcda - cdaa + aaaa \qquad\qquad a = b + c + d
\end{array}
$$

If $bc + bd + cd = bf + cf + df$ the second degree term (in aa) is removed,

for then:
$$\frac{bc + bd + cd}{b + c + d} = f$$

and by reduction the equation becomes:

$$
\begin{array}{lll}
+\,bbccd\ = & -\,bbcca & +\,bbaaa \\
+\,bbcdd & -\,bbdda & +\,ccaaa \\
+\,bccdd & -\,ccdda & +\,ddaaa \\
\overline{b + c + d} & -\,bcdda & +\,bcaaa \\
& -\,bccda & +\,bdaaa \\
& -\,bbcda & +\,cdaaa \quad +\,aaaa \\
& \overline{b + c + d}\ \ \overline{b + c + d}
\end{array}
$$

$$a = \frac{bc + bd + cd}{b + c + d}$$

If $bcd = bcf + bdf + cdf$ the first degree term (in a) is removed,

for then:
$$\frac{bcd}{bc + bd + cd} = f$$

and by reduction the equation becomes:

$$
\begin{array}{lll}
\dfrac{bbccdd}{bc + bd + cd}\ = & +\,bbccaa & +\,bbcaaa \\
& +\,bbddaa & +\,bbdaaa \\
& +\,ccddaa & +\,bccaaa \\
& +\,bcddaa & +\,ccdaaa \\
& +\,bccdaa & +\,bddaaa \\
& +\,bbcdaa & +\,cddaaa \\
& \overline{bc + bd + cd} & +\,2bcdaaa \quad +\,aaaa \\
& & \overline{bc + bd + cd}
\end{array}
$$

$$a = \frac{bcd}{bc + bd + cd}$$

$$aaaa + 8aaa + 18aa = 27 = +18aa - 8aaa + aaaa$$

$$\frac{bcd}{bc + bd + cd} = \frac{27}{9 + 9 + 9}$$

$a = 1$

$a = 3 = b$

$a = 3 = c$

$a = 3 = d$

d.9) *On the generation of canonical equations*

$$
\left.\begin{array}{l} a - b \\ a - c \\ a + d \\ a + f \end{array}\right|
\begin{array}{l}
= aaaa - baaa \\
\quad - caaa + bcaa \\
\quad + daaa - bdaa \\
\quad + faaa \ - cdaa + bcda \\
\qquad\quad - bfaa + bcfa \\
\qquad\quad - cfaa - bdfa \\
\qquad\quad + dfaa - cdfa + bcdf = 0000
\end{array}
$$

therefore: $bcdf = -\ bcda - bcaa$
$$
\begin{array}{ll}
\quad - bcfa \ + bdaa & \\
\quad + bdfa \ + cdaa + baaa & \\
\quad + cdfa \ + bfaa + caaa & \\
\qquad\quad + cfaa - daaa & a = b \\
\qquad\quad - dfaa - faaa - aaaa & a = c
\end{array}
$$

From this equation arise the three following trinomial equations and also three binomial equations.

If $b + c = d + f$ the third degree term (in *aaa*) is removed,
for then $b + c - d = f$ and by reduction the equation becomes:

$$
\begin{array}{ll}
+\ bbcd = & -\ bbca \\
+\ bccd & -\ bcca - bdaa \\
-\ bcdd & -\ bdda - cdaa \\
& -\ cdda + bbaa \\
& +\ bbda + bcaa \\
& +\ ccda + ccaa & \qquad a = b \\
& +\ 2bcda + ddaa \ - aaaa & \qquad a = c
\end{array}
$$

If $bc + df = bd + cd + bf + cf$ the second degree term (in aa) is removed, for then, first, $bc - bd - cd = bf + cf - df$

whence: $bc - bd - cd = f$
$$\overline{b + c - d}$$

and by reduction the equation becomes:

$$
\begin{array}{lll}
+\, bbccd = & -\, bbcca & +\, bbaaa \\
-\, bbcdd & -\, bbdda & +\, ccaaa \\
-\, bccdd & -\, bcdda & +\, ddaaa \\
\overline{b + c - d} & -\, ccdda & +\, bcaaa \\
 & +\, bbcda & -\, bdaaa \\
 & +\, bccda & -\, cdaaa \quad -\, aaaa \\
 & \overline{b + c - d} \;\; \overline{b + c - d} \\
\end{array}
\qquad
\begin{array}{l}
a = b \\
a = c
\end{array}
$$

and, second, $df - bf - cf = bd + cd - bc$

whence: $bd + cd - bc = f$
$$\overline{d - b - c}$$

and by reduction the equation becomes:

$$
\begin{array}{lll}
+\, bbcdd = & +\, bbcca & -\, bbaaa \\
+\, bccdd & +\, bbdda & -\, bcaaa \\
-\, bbccd & +\, bcdda & -\, ccaaa \\
\overline{d - b - c} & +\, ccdda & -\, ddaaa \\
 & -\, bbcda & +\, bdaaa \\
 & -\, bccda & +\, cdaaa \quad -\, aaaa \\
 & \overline{d - b - c} \;\; \overline{d - b - c} \\
\end{array}
\qquad
\begin{array}{l}
a = b \\
a = c
\end{array}
$$

d.10)   *On the generation of canonical equations*

If $bcd + bcf = bdf + cdf$ the first degree term (in a) is removed,

$\quad bcd = bdf + cdf - bcf$

so $\dfrac{bcd}{bd + cd - bc} = f$ and by reduction the equation becomes:

$$\frac{bbccdd}{bd + cd - bc} = \frac{\begin{array}{ll} + bbccaa & - bbcaaa \\ + bbddaa & - bccaaa \\ + bcddaa & - bddaaa \\ + ccddaa & - cddaaa \\ - bbcdaa & + bbdaaa \\ - bccdaa & + ccdaaa \\ \hline bd + cd - bc & + 2bcdaaa \quad - aaaa \end{array}}{bd + cd - bc}$$

$a = b$

$a = c$

On the origin of binomial equations from the previous primary equation (d.9). First that produced by removing the second and third degree terms (in aa and aaa).

It is required that: $b + c = d + f$

so: $b + c - d = f$

[and that][5] $bc + df = bd + cd + bf + cf$

then: $bc + (bd + cd - dd) = bd + cd + (bb + bc - bd) + (bc + cc - cd)$

therefore: $bb + bc + cc = bd + cd - dd$

which has two roots.

Therefore one root $d = \dfrac{+b + c}{2} - \sqrt{\dfrac{bb - 2bc + cc}{4} + \dfrac{-4bb - 4bc - 4cc}{4}}$

that is: $d = \dfrac{b + c}{2} - \sqrt{\dfrac{-3bb - 2bc - 3cc}{4}}$

and because $\quad b + c - d = f$

we will have: $\quad f = b + c - b - c + \dfrac{}{2} \quad \sqrt{\dfrac{-3bb - 2bc - 3cc}{4}}$

that is: $\quad f = \dfrac{b + c}{2} + \sqrt{\dfrac{-3bb - 2bc - 3cc}{4}}$

and this is the other root of the equation.

Because $\quad d = \dfrac{b + c}{2} - \sqrt{\dfrac{-3bb - 2bc - 3cc}{4}}$

and $\quad f = \dfrac{b + c}{2} + \sqrt{\dfrac{-3bb - 2bc - 3cc}{4}}$

therefore $\quad d + f = b + c$ as was required

and: $\quad df = \dfrac{bb + 2bc + cc}{4} + \dfrac{3bb + 2bc + 3cc}{4}$

that is: $\quad df = \dfrac{4bb + 4bc + 4cc}{4}$

therefore $\quad df = bb + bc + cc$, the given constant term of the equation that arises.

Because $\quad d + f = b + c$
and $\quad df = bb + bc + cc$
the coefficients of the first degree term are reduced thus:

$$- bcd - bcf = -bbc - bcc$$
$$+ bdf = bbb + bbc + bcc$$
$$+ cdf = bbc + bcc + ccc$$

and:

$$bcdf = bbbc + bbcc + bccc$$

Therefore the sought binomial equation is:

$$
\begin{array}{ll}
\begin{aligned}
bbbc &= + bbba \\
+bbcc &\quad + bbca \\
+bccc &\quad + bcca \\
&\quad + ccca - aaaa
\end{aligned}
&
\begin{aligned}
&a = b \\[1em]
&a = c
\end{aligned}
\end{array}
$$

---

d.II) *On the generation of canonical equations*

Second, the origin of the binomial equation formed by the removal of the first and second degree terms (in a and aa).

It is required that: $bc + df = bd + cd + bf + cf$
therefore: (1) $bc - bd - cd = bf + cf - df$

whence: $\dfrac{bc - bd - cd}{b + c - d} = f$

[and that] $bcd + bcf = bdf + cdf$

then: $\dfrac{bbcd + bccd - bcdd + bbcc - bbcd - bccd}{b + c - d} \quad \dfrac{}{b + c - d}$

$= \dfrac{bbcd - bbdd - bcdd + bccd - bcdd - ccdd}{b + c - d} \quad \dfrac{}{b + c - d}$

therefore: $bbcc = + bbcd - bbdd$
$\qquad\qquad\quad + bccd - bcdd$
$\qquad\qquad\qquad\quad - ccdd$

therefore: $\dfrac{bbcc}{bb + bc + cc} = \dfrac{bbcd}{\quad + bccd \quad} - dd$
$\qquad\qquad\qquad\qquad\qquad \dfrac{+ bccd}{bb + bc + cc}$

a quadratic equation with two roots.

Therefore:[6] 1st $d = \dfrac{bbc + bcc}{2bb + 2bc + 2cc} - \sqrt{}$

$\qquad\qquad 2^{nd}$ $d = \dfrac{bbc + bcc}{2bb + 2bc + 2cc} + \sqrt{}$

Or: (2) $bc - bf - cf = bd + cd + fd$

whence: $\dfrac{bc - bf - cf}{b + c - f} = d$

Then by a similar argument to that above, or everywhere interchanging d and f, we arrive at the same quadratic equation with two roots:

$$\frac{bbcc}{bb + bc + cc} = \frac{bbcf}{bb + bc + cc} \begin{array}{c} \\ + bccf \\ \end{array} - ff$$

Therefore: 1st $f = \dfrac{bbc + bcc}{2bb + 2bc + 2cc} \quad - \sqrt{}$

2nd $f = \dfrac{bbc + bcc}{2bb + 2bc + 2cc} \quad + \sqrt{}$

the same roots as above.

Therefore if d and f are different roots, as required, $d+f$ is the sum of a binome and its residual.

that is: $d + f = \dfrac{bbc + bcc}{bb + bc + cc}$

and $df = \dfrac{bbcc}{bb + bc + cc}$ as expected, the given constant term,

for in quadratic equations where the square term is negative, the product of the two roots is equal to the constant.

Then the coefficients of the third degree term (in aaa) are reduced thus:

$-d - f = \dfrac{- bbc - bcc}{bb + bc + cc}$

$b = \dfrac{+ bbb + bbc + bcc}{bb + bc + cc}$

$c = \dfrac{+ bbc + bcc + ccc}{bb + bc + cc}$

and the constant term thus:

$bcdf = \dfrac{bbbccc}{bb + bc + cc}$

Therefore the sought binomial equation is:

$$\dfrac{bbbccc}{bb + bc + cc} \quad = \quad \begin{array}{l} + bbbaaa \\ + bbcaaa \\ + bccaaa \\ \underline{+ cccaaa} \\ bb + bc + cc \end{array} \quad - aaaa \qquad \begin{array}{l} a = b \\[4pt] a = c \end{array}$$

This equation is correct and is demonstrated in (c.14).

Therefore what was supposed in this investigation is also correct.

Add MS 6783 f. 172

d.12)   *On the generation of canonical equations*

Third and last, the origin of the binomial equation formed by the removal of the first and third degree terms (in a and aaa).

It is required that: $d + f = b + c$
that is: $d = b + c - f$

[and that] $bcd + bcf = bdf + cdf$

then: $(bbc + bcc - bcf) + bcf = (bbf + bcf - bff) + (bcf + ccf - cff)$

$$\begin{array}{l} + bbc = \quad + bbf \\ \underline{+ bcc \quad\quad + 2bcf} \\ b + c \quad\quad + ccf \quad - ff \\ \quad\quad\quad\quad \overline{b + c} \end{array}$$

a quadratic equation with two roots.

$$1^{\text{st}} \; f = \dfrac{bb + 2bc + cc - \sqrt{}}{2b + 2c}$$

$$2^{\text{nd}} \; f = \dfrac{bb + 2bc + cc + \sqrt{}}{2b + 2c}$$

Because $d = b + c - f$

therefore: 1st $d = b + c - \dfrac{bb - 2bc - cc + \sqrt{}}{2b + 2c}$

that is: 1st $d = \dfrac{2bb + 4bc + 2cc}{2b + 2c} - \dfrac{bb - 2bc - cc + \sqrt{}}{2b + 2c}$

that is: 1st $d = \dfrac{bb + 2bc + cc + \sqrt{}}{2b + 2c}$

therefore: 2nd $d = \dfrac{bb + 2bc + cc - \sqrt{}}{2b + 2c}$

Therefore $d + f =$ the sum of a binome and its residual,

that is: $d + f = \dfrac{bb + 2bc + cc}{b + c}$

and $df = \dfrac{bbcc}{b + c}$, as expected, the given constant term of the quadratic equation.

Then the coefficients of the second degree term (in aa) are reduced thus:

$- bc = - \dfrac{bbc - bcc}{b + c}$

$- df = - \dfrac{bbc - bcc}{b + c}$

$+ bd + bf = + \dfrac{bbb + 2bbc + bcc}{b + c}$

$+ cd + cf = + \dfrac{bbb + 2bbc + bcc}{b + c}$

and the constant term thus:

$bcdf = \dfrac{bbbcc + bbccc}{b + c}$

Therefore the sought binomial equation is:

$\dfrac{bbbcc + bbccc}{b + c} = \dfrac{+ bbbaa + bbcaa + bccaa + cccaa - aaaa}{b + c}$

$a = b$

$a = c$

Alternatively, from part of the preceding work:
since

$$+ bbc = + bbf$$
$$\frac{+ bcc \qquad + 2bcf}{b + c \qquad + ccf \ - ff}$$
$$\overline{\qquad\quad b + c}$$

we will have $bc = bf + cf - ff$

a quadratic equation with two roots.

Therefore: $1^{st} f = \dfrac{b + c}{2} - \sqrt{\dfrac{bb + 2bc + cc}{4} - \dfrac{4bc}{4}}$

that is: $\quad 1^{st} f = \dfrac{b + c}{2} - \sqrt{\dfrac{bb - 2bc + cc}{4}}$

that is: $\quad 1^{st} f = \dfrac{b + c}{2} - \dfrac{b + c}{2}$

therefore: $\quad 1^{st} f = c$

Therefore: $2^{nd} f = \dfrac{b + c}{2} + \sqrt{\dfrac{bb + 2bc + cc}{4} - \dfrac{4bc}{4}}$

$2^{nd} f = \dfrac{b + c}{2} + \sqrt{\dfrac{bb - 2bc + cc}{4}}$

$2^{nd} f = \dfrac{b + c}{2} + \dfrac{b - c}{2}$

therefore: $2^{nd} f = b$

Since $d = b + c - f$

therefore: $1^{st} d = b + c - c = b$

therefore: $2^{nd} d = b + c - b = c$

therefore $d + f = b + c$, the binomium plus its residual, and $df = bc$ the given constant term.

Then the coefficients of the second degree term (in aa) are reduced thus:

$$- bc = - bc$$
$$- df = - bc$$
$$+ bd + bf = bb + bc$$
$$+ cd + cf = bc + cc$$

and the constant term thus:

$$bcdf = bbcc$$

Therefore the sought binomial equation is:

$$bbcc = \quad bbaa \qquad\qquad\qquad\qquad a = b$$
$$\quad\quad + ccaa - aaaa \qquad\qquad\qquad a = c$$

The preceding alternative equation may be reduced to the same as this. Either form is useful in its own time and place.

Add MS 6783 f. 171

$d.13$)   *On the generation of canonical equations*

On reciprocal equations

$$\left.\begin{array}{r} a - b \\ aaa - cdf \end{array}\right| = aaaa - baaa - cdfa + bcdf = 0000$$

therefore: $bcdf = + cdfa + baaa - aaaa$ $a = b$
                                            $a = \sqrt[3]{cdf}$

and: $bccc = + ccca + baaa - aaaa$ $a = b$
                                            $a = c$

$$\left.\begin{array}{r} a + b \\ aaa - cdf \end{array}\right| = aaaa + baaa - cdfa - bcdf = 0000$$

therefore: $bcdf = - cdfa + baaa + aaaa$ $a = \sqrt[3]{cdf}$

and: $bccc = - ccca + baaa - aaaa$ $a = c$

$$a - b \left| \begin{array}{l} \\ aaa + cdf \end{array} \right. = aaaa - baaa + cdfa - bcdf = \text{oooo}$$

therefore: $bcdf = + cdfa - baaa + aaaa$ $\qquad a = b$

and: $bccc = + ccca - baaa + aaaa$ $\qquad a = b$

The root is known here without solving, as in cubic equations under similar conditions (*d*.6).

Add MS 6783 f. 156

d.13.2)[7] $\qquad\qquad$ *On the generation of canonical equations*

$$\begin{array}{l} b - a \\ c - a \\ df - aa \end{array} \left| \begin{array}{l} = bcdf - bdfa + dfaa + baaa \\ \quad - cdfa - bcaa + caaa - aaaa = \text{oooo} \end{array} \right.$$

Therefore: $bcdf = + bdfa + bcaa - baaa$ $\qquad a = b$

$\qquad\qquad\quad + cdfa - dfaa - caaa + aaaa$ $\qquad a = c$

$\qquad\qquad\qquad\qquad\qquad\qquad\qquad\qquad\qquad a = \sqrt{df}$

$$\begin{array}{l} b - a \\ c - a \\ df + aa \end{array} \left| \begin{array}{l} = bcdf - bdfa + dfaa - baaa \\ \quad - cdfa + bcaa - caaa + aaaa = \text{oooo} \end{array} \right.$$

Therefore: $bcdf = + bdfa - bcaa + baaa$ $\qquad a = b$

$\qquad\qquad\quad + cdfa - dfaa + caaa - aaaa$ $\qquad a = c$

$\qquad\qquad\qquad\qquad\qquad\qquad\qquad\qquad\qquad aa = - df$

$\qquad\qquad\qquad\qquad\qquad\qquad\qquad\qquad\qquad a = \sqrt{- df}$

$$\begin{array}{l} b + a \\ c + a \\ df - aa \end{array} \left| \begin{array}{l} = bcdf + bdfa + dfaa - baaa \\ \quad + cdfa - bcaa - caaa - aaaa = \text{oooo} \end{array} \right.$$

Therefore: $bcdf = - bdfa + bcaa + baaa$ $\qquad a = \sqrt{df}$

$\qquad\qquad\quad - cdfa - dfaa + caaa + aaaa$ $\qquad a = - b$

$\qquad\qquad\qquad\qquad\qquad\qquad\qquad\qquad\qquad a = - c$

$$\left.\begin{array}{c} b - a \\ c + a \\ df + aa \end{array}\right| = bcdf + bdfa - dfaa + baaa$$
$$- cdfa + bcaa - caaa - aaaa = 0000$$

Therefore: $\quad bcdf = -\,bdfa - bcaa - baaa \qquad\qquad a = b$
$$+\, cdfa + dfaa + caaa + aaaa \qquad\qquad a = -c$$
$$aa = -\,df$$
$$a = \sqrt{-\,df}$$

$$\left.\begin{array}{c} b + a \\ c - a \\ df - aa \end{array}\right| = bcdf - bdfa - dfaa + baaa$$
$$+ cdfa - bcaa - caaa + aaaa = 0000$$

Therefore: $\quad bcdf = +\,bdfa + bcaa - baaa \qquad\qquad a = c$
$$-\, cdfa + dfaa + caaa - aaaa \qquad\qquad a = \sqrt{df}$$
$$a = -\,b$$

$$\left.\begin{array}{c} bc - aa \\ df - aa \end{array}\right| = bcdf - dfaa$$
$$- bcaa + aaaa = 0000$$

Therefore: $\quad bcdf = +\,bcaa \qquad\qquad aa = bc$
$$+\, dfaa - aaaa \qquad\qquad aa = df$$

$$\left.\begin{array}{c} bc - aa \\ df + aa \end{array}\right| = bcdf - dfaa$$
$$+ bcaa - aaaa = 0000$$

Therefore: $\quad bcdf = -\,bcaa \qquad\qquad aa = bc$
$$+\, dfaa - aaaa \qquad\qquad aa = -\,df$$

d.14)   *On the generation of canonical equations*

A collection of various equations of this kind, that the generation of other
higher degree terms may be more easily apparent

d.1) $bc = + ba$ $a = b$
           $+ ca - aa$ $a = c$

d.3) $bbc = + bba$
       $+ bcc$ $+ bca$ $a = b$
           $+ cca - aaa$ $a = c$

d.10) $bbbc = + bbba$
       $+ bbcc$ $+ bbca$
       $+ bccc$ $+ bcca$ $a = b$
           $+ ccca - aaaa$ $a = c$

   $bbbbc = + bbbba$
$+ bbbcc$ $+ bbbca$
$+ bbccc$ $+ bbcca$
$+ bcccc$ $+ bccca$ $a = b$
       $+ ccccа - aaaaa$ $a = c$

And so on for the rest as needed.

d.3) $\dfrac{bbcc}{b+c}$ $= + bbaa$
           $+ bcaa$
           $\dfrac{+ ccaa}{b+c}$ $- aaa$ $a = b$
                                            $a = c$

d.12) $bbbcc = + bbbaa$
       $\dfrac{+ bbccc}{b+c}$ $+ bbcaa$
                       $+ bccaa$
           $\dfrac{+ cccaa}{b+c}$ $- aaaa$ $a = b$
                                              $a = c$

$$\frac{\begin{aligned}bbbbcc &= \\ + bbbccc & \\ + bbcccc & \\ \hline b + c & \end{aligned}}{} \quad \begin{aligned} &+ bbbbaa \\ &+ bbbcaa \\ &+ bbccaa \\ &+ bcccaa \\ &\underline{+ ccccaa \quad - aaaaa} \\ &\quad\quad b + c \end{aligned} \qquad \begin{aligned} a &= b \\ a &= c \end{aligned}$$

And so on for the rest as needed.

$d.11)$

$$\frac{bbbccc}{bb + bc + cc} = \begin{aligned} &+ bbbaa \\ &+ bbcaaa \\ &+ bccaaa \\ &\underline{+ cccaaa \quad - aaaa} \\ &\quad bb + bc + cc \end{aligned} \qquad \begin{aligned} a &= b \\ a &= c \end{aligned}$$

$$\frac{\begin{aligned}bbbbccc &= \\ + bbbcccc & \\ \hline bb + bc + cc & \end{aligned}}{} \quad \begin{aligned} &+ bbbbaaa \\ &+ bbbcaaa \\ &+ bbccaaa \\ &+ bcccaaa \\ &\underline{+ ccccaaa \quad - aaaaa} \\ &\quad\quad bb + bc + cc \end{aligned} \qquad \begin{aligned} a &= b \\ a &= c \end{aligned}$$

And so on for the rest as needed.

$$\frac{bbbbcccc}{bbb + bbc + bcc + ccc} = \begin{aligned} &bbbbaaaa \\ &+ bbbcaaaa \\ &+ bbccaaaa \\ &+ bcccaaaa \\ &\underline{+ ccccaaaa \quad - aaaaa} \\ &\quad bbb + bbc + bcc + ccc \end{aligned} \qquad \begin{aligned} a &= b \\ a &= c \end{aligned}$$

And so on for the rest as needed.

Add MS 6783 f. 169

d.15)   *On the generation of canonical equations*

Another collection and series of canonicals

d.2) $bcd = + bca - baa$ $a = b$
              $+ bda - caa$ $a = c$
              $+ cda - daa + aaa$ $a = d$

d.7) $bbcd =$ $+ bbca - bbaa$
       $+ cbcd$ $+ bbda - ccaa$
       $+ dbcd$ $+ ccba - ddaa$
                   $+ ccda - bcaa$
                   $+ ddba - bdaa$ $a = b$
                   $+ ddca - cdaa + aaaa$ $a = c$
              $+ 2bcda$ $a = d$

d.7) $bcbcd =$ $+ bbcca - bbaaa$
       $+ bdbcd$ $+ bbdda - ccaaa$
       $\dfrac{+ cdbcd}{b + c + d}$ $+ ccdda - ddaaa$
                   $+ bbcda - bcaaa$
                   $+ cbcda - bdaaa$ $a = b$
                   $\dfrac{+ dbcda}{b + c + d}$ $\dfrac{- cdaaa + aaaa}{b + c + d}$ $a = c$
                                                      $a = d$

d.7) $\dfrac{bcdbcd}{bc + bd + cd} =$ $+ bbccaa - bbcaaa$
                   $+ bbddaa - bbdaaa$
                   $+ ccddaa - ccbaaa$
                   $+ bbcdaa - ccdaaa$
                   $+ cbcdaa - ddbaaa$
                   $\dfrac{+ dbcdaa}{bc + bd + cd}$ $- ddcaaa$ $a = b$
                                   $\dfrac{- 2bcdaaa + aaaa}{bc + bd + cd}$ $a = c$
                                                                              $a = d$

162

$$
\begin{aligned}
bbc\,bcd = &+ bbbcca && - bbbaaa \\
+ bbd\,bcd \quad &+ bbbdda && - cccaaa \\
+ ccb\,bcd \quad &+ cccbba && - dddaaa \\
+ ccd\,bcd \quad &+ bbbcda && - bbcaaa \\
+ ddb\,bcd \quad &+ cccdda && - bbdaa \\
+ ddc\,bcd \quad &+ cccbda && - ccbaa \\
+ 2bcd\,bcd \quad &+ dddbba && - ccdaa \\
\overline{b+c+d} \quad &+ dddcca && - ddbaa \\
&+ dddbca && - ddcaa \\
&+ 2bcbcd && - bcdaa \quad + aaaaa \\
&+ 2bdbcd && \overline{b+c+d} \\
&+ 2cdbcd && \\
&\overline{b+c+d} &&
\end{aligned}
$$

$$
\begin{aligned}
a &= b \\
a &= c \\
a &= d
\end{aligned}
$$

And so on for others as needed.

Add MS 6783 f. 168

d.16)   *On the generation of canonical equations*

Another collection and series of canonicals

$$
\begin{array}{lll}
& b = a & a = b \\
d.1) & bc = + ba & a = b \\
& + ca - aa & a = c \\
d.2) & bcd = + bca & \\
& + bda - baa & a = b \\
& + cda - caa & a = c \\
&   - daa + aaa & a = d \\
d.7) & bcdf = + bcda - bcaa & \\
& + bcfa - bdaa & \\
& + bdfa - cdaa + baaa & a = b \\
& + cdfa - bfaa + caaa & a = c \\
&   - cfaa + daaa & a = d \\
&   - dfaa + faaa - aaaa & a = f
\end{array}
$$

$$bcdfg = + bcdfa - bcdaa + bcaaa$$
$$+ bcdga - bcfaa + bdaaa$$
$$+ bcfga - bdfaa + cdaaa$$
$$+ bdfga - cdfaa + bfaaa$$
$$+ cdfga - bcgaa + cfaaa$$

$- bdgaa + dfaaa - baaaa$	$a = b$
$- cdgaa + bgaaa - caaaa$	$a = c$
$- bfgaa + cgaaa - daaaa$	$a = d$
$- cfgaa + dgaaa - faaaa$	$a = f$
$- dfgaa + fgaaa - gaaaa + aaaaa$	$a = g$

Etc.

$1 = a$	$a = 1$
$2 = +3a - aa$	$a = 1,2$
$6 = +11a - 6aa + aaa$	$a = 1,2,3$
$24 = +50a - 35aa + 10aaa - aaaa$	$a = 1,2,3,4$
$120 = +274a - 225aa + 85aaa - 15aaaa + aaaaa$	$a = 1,2,3,4,5$

Etc.

d.17)   *On the generation of canonical equations*

Appendix 1. On finding numbers fitting certain conditions[8]

d.5) To find numbers of the form: $\dfrac{bc}{b+c}$

$b + c : b = c : \dfrac{bc}{b+c}$

therefore: $\quad \left. b \;\middle|\; +c \;\middle|\; : b \;\middle|\; = c \;\middle|\; : b \right.$
$\qquad\quad b+c \quad b+c \quad b+c \quad b+c \quad c$

Therefore the sought quantity in letters will be:

$$\left. \begin{array}{c} b \\ b+c \\ c \\ b+c \\ \hline b \;\middle|\; +c \\ b+c \;\middle|\; b+c \end{array} \right| = \left. \begin{array}{c} bc \\ b+c \\ b+c \\ \hline b+c \\ b+c \\ \hline b+c \end{array} \right| = bc$$

Let $b = 1, c = 1$, therefore: $\qquad \left. \begin{array}{c} 2 \\ 2 \\ \hline 2+2 \end{array} \right| = \dfrac{4}{4} = 1$

Let $b = 1, c = 2$, therefore: $\qquad \left. \begin{array}{c} 3 \\ 6 \\ \hline 3+6 \end{array} \right| = \dfrac{18}{9} = 2$

d.8) To find numbers of the form: $\dfrac{bc + bd + cd}{b + c + d} = f$

The sought quantity in letters will be:

$$
\begin{array}{l|l|l|l}
b & +b & +c & = bc + bd + cd \\
b+c+d & b+c+d & b+c+d & b+c+d \\
c & d & d & b+c+d \\
b+c+d & b+c+d & b+c+d & b+c+d \\
b & +c & d & b+c+d \\
b+c+d & b+c+d & b+c+d & b+c+d
\end{array} \quad = bc + bd + cd
$$

Let $b = 1, c = 1, d = 1$, therefore: $\dfrac{9 + 9 + 9}{3 + 3 + 3} = \dfrac{27}{9} = 3$

Let $b = 1, c = 2, d = 3$, therefore: $\dfrac{\begin{array}{r|r|r} 6 & +\ 6 & +12 \\ 12 & +18 & +\ 8 \end{array}}{6 + 12 + 8} = \dfrac{396}{30} = 11$

d.18)   *On the generation of canonical equations*

Appendix 1

d.8) To find numbers of the form: $\dfrac{bcd}{bc + bd + cd} = f$

The sought quantity in letters will be:

$$
\begin{array}{c|c|c}
b & = bcd & = bcd \\
bc + bd + cd & bc + bd + cd & \\
c & bc + bd + cd & \\
bc + bd + cd & bc + bd + cd & \\
d & \overline{bc + bd + cd} & \\
bc + bd + cd & bc + bd + cd & \\
 & bc + bd + cd &
\end{array}
$$

$$
\begin{array}{c|c|c}
bc & + \; bd & + \; cd \\
bc + bd + cd & bc + bd + cd & bc + bd + cd \\
bc + bd + cd & bc + bd + cd & bc + bd + cd
\end{array}
$$

Let $b = 1, c = 1, d = 1$, therefore:

$$
\begin{array}{c|c}
\left.\begin{array}{c} 3 \\ 3 \\ 3 \end{array}\right. & = \dfrac{27}{27} = 1 \\
\hline
9 + 9 + 9 &
\end{array}
$$

Let $b = 1, c = 2, d = 3$, therefore:

$$
\begin{array}{c|c}
\left.\begin{array}{c} 11 \\ 22 \\ 33 \end{array}\right. & = \dfrac{7986}{1331} = 6
\end{array}
$$

$$
\begin{array}{c|c|c}
11 & + \; 11 & + \; 22 \\
22 & 33 & 33
\end{array}
$$

To find numbers of the form: $\dfrac{bcd}{b+c+d}$

The sought quantity in letters will be:

$$
\begin{array}{c|c|c}
b & = & bcd & = & bcd \\
b+c+d & & b+c+d & & b+c+d \\
c & & b+c+d & \\
b+c+d & & b+c+d & \\
d & & b+c+d & \\
b+c+d & & b+c+d & \\
\hline
\begin{array}{c|c|c} b & +c & +d \\ b+c+d & b+c+d & b+c+d \end{array}
\end{array}
$$

Let $b = 1, c = 1, d = 1$, therefore:

$$
\begin{array}{c}
3 \\
3 \\
3 \\
\hline
3+3+3
\end{array}
\bigg| = \dfrac{27}{9} = 3
$$

Let $b = 1, c = 2, d = 3$, therefore:

$$
\begin{array}{c}
6 \\
12 \\
18 \\
\hline
6+12+18
\end{array}
\bigg| = \dfrac{1296}{36} = 36
$$

d.19)   *On the generation of canonical equations*

Appendix I

To find numbers of the form: $\dfrac{bb + cc}{b + c}$

The sought quantity in letters will be:

$$\begin{array}{|c|c|}\hline b & +c \\ b+c & b+c \\ b & c \\ b+c & b+c \\\hline b & +c \\ b+c & b+c \\\hline\end{array}\ \begin{array}{|c|}\hline = bb+cc \\ b+c \\\hline b+c \\ b+c \\\hline b+c \\ b+c \\\hline\end{array}\ = bb + cc$$

Let $b = 2, c = 3$, therefore: $\begin{array}{|c|c|}\hline 10 & +15 \\ 10 & 15 \\\hline 10 & +15 \\\hline\end{array}\ \begin{array}{|c|}\hline = 4+9 \\ 2+3 \\\hline 2+3 \\ 2+3 \\\hline 2+3 \\\hline\end{array}\ = 4 + 9$

To find numbers of the form: $\dfrac{bbb + ccc}{b + c}$

The sought quantity in letters will be:

$$\begin{array}{|c|c|}\hline b & +c \\ b+c & b+c \\ b & c \\ b+c & b+c \\ b & c \\ b+c & b+c \\\hline b & +c \\ b+c & b+c \\\hline\end{array}\ \begin{array}{|c|}\hline = bbb+ccc \\ b+c \\\hline b+c \\ b+c \\\hline b+c \\ b+c \\\hline b+c \\\hline\end{array}\ \begin{array}{|c|}\hline = bbb+ccc \\ b+c \\\hline\end{array}$$

Although the foregoing formula is true, however:

$$\frac{bbb + ccc}{b + c} = bb - bc + cc$$

and this is therefore by its nature such a number in letters.

For:

$$bb - bc + cc = \left.\begin{array}{c} bb - bc + cc \\ \underline{\hspace{1cm} b + c} \\ b + c \end{array}\right| = \begin{array}{c} bbb - bbc + bcc \\ + bbc - bcc + ccc \\ \underline{\hspace{2cm}} \\ b + c \end{array} = \frac{bbb + ccc}{b + c}$$

And this is the rule for the sign, of which use will be made at some time or other in what follows. And others as above.

Thus:

$$\frac{bbb - ccc}{b - c} = bb + bc + cc$$

Add MS 6783 f. 164

d.20)   *On the generation of canonical equations*

Appendix 2. On the multiplication of roots

Let the equation to be solved be:
$$daaa + bbaa + ccca = xxxz$$

then, by parabolismus:
$$aaa + bbaa + ccca = xxxz$$
$$\frac{}{d} \quad \frac{}{d} \quad \frac{}{d}$$

If in numbers *d* does not divide any of the above numbers, the terms may be multiplied by a series of quantities in continued proportion, thus:

$$\begin{array}{cccc} 1 & d & dd & ddd \end{array}$$

therefore:
$$aaa + bbaa + dccca \quad ddxxxz$$

I say that in this equation the root is *da*, where *a* is the root of the original equation.

Let *a* be [replaced by] *da*

therefore: $$dadada + bbdada + dcccda = ddxxxz$$

that is: $$dddaaa + bbddaa + cccdda = ddxxxz$$

therefore: $$aaa + \frac{bbddaa}{ddd} + \frac{cccdda}{ddd} = \frac{ddxxxz}{ddd}$$

therefore: $$aaa + \frac{bbaa}{d} + \frac{ccca}{d} = \frac{xxxz}{d}$$

and it is so. Therefore the proposition is true.

An example in numbers:

Let: $3aaa + 7aa + 5a = 62$ $a = 2$

therefore: $$\frac{3aaa}{3} + \frac{7aa}{3} + \frac{5a}{3} = \frac{62}{3}$$ $a = 2$

that is: $$aaa + \frac{7aa}{3} + \frac{5a}{3} = \frac{62}{3}$$ $a = 2$

then multiplying by: $\begin{array}{cc} 1 & 3 \end{array}$ $\begin{array}{c|c} 3 & 3 \\ 3 & 3 \\ & 3 \end{array}$

therefore: $aaa + 7aa + 15a = 558$ $a = 2 \left|\begin{array}{c} \\ 3 \end{array}\right. = 6$

for: $\left.\begin{array}{c} 6 \\ 6 \\ 6 \end{array}\right| + \left.\begin{array}{c} 6 \\ 6 \\ 7 \end{array}\right| + \left.\begin{array}{c} 6 \\ 15 \end{array}\right| = 558$

therefore: $\dfrac{6}{3} = 2$ is the root *a* of the original equation.

d.21)   *On the generation of canonical equations*

Appendix 2. On the multiplication of roots

In the given equation $\qquad\qquad aaa - baa + fga = xxz$

let the roots be multiplied by d.

There will be proportionals: $\qquad$ I $\qquad d \qquad dd \qquad ddd$

therefore: $\qquad aaa - dbaa + ddfga = dddxxz$

I say that in this equation the root is da, where a is the root of the original equation.

Let a be [replaced by] da

therefore: $\quad dadada - dbdada + ddfgda = dddxxz$

that is: $\qquad dddaaa - dddbaa + dddfga = dddxxz$

therefore: $\qquad aaa - \dfrac{dddbaa}{ddd} + \dfrac{dddfga}{ddd} = \dfrac{dddxxz}{ddd}$

therefore: $\qquad\qquad aaa - baa + fga = xxz$

and it is so. Therefore the proposition is true.

An example in numbers

In the given equation let any root be doubled.

$aaa - 6aa + 11a = 6$ $\qquad\qquad\qquad\qquad a =$ I, 2, 3

$$\begin{array}{ccccc} \text{I} & 2 & 2 & & 2 \\ & & \underline{2} & & 2 \\ & & & & \underline{2} \\ & & & & 2 \end{array}$$

therefore: $\quad aaa - 12aa + 44a = 48 \qquad\qquad a = 2, 4, 6$

Another example in numbers

Multiply the roots by 10

$aaa + 7a = 22$ $\qquad\qquad$ $a = 2$

that is: $\qquad aaa + 0aa + 7a = 22$ $\qquad a = 2$

$\qquad\qquad$ I $\qquad$ IO $\quad$ IOO $\quad$ IOOO

therefore: $aaa + 700a = 22000$ $\qquad a = 20$

$\qquad\qquad aaa + 3aa = 20$ $\qquad a = 2$

that is: $\qquad aaa + 3aa + 0a = 20$

$\qquad\qquad$ I $\qquad$ IO $\quad$ IOO $\quad$ IOOO

therefore: $aaa + 30aa = 20000$ $\qquad a = 20$

$aa - 7a = 8$ $\qquad\qquad a = 8$

I $\quad$ IO $\quad$ IOO

$aa - 70a = 800$ $\qquad a = 80$

Corollary

From the multiplication of the roots by ten, or one hundred, etc. it becomes apparent what fractions (tenths or hundredths, etc.) are to be had in the solution. This is clear from the proper addition of zeros and points, as the powers and coefficients require.

And so on for all other equations.

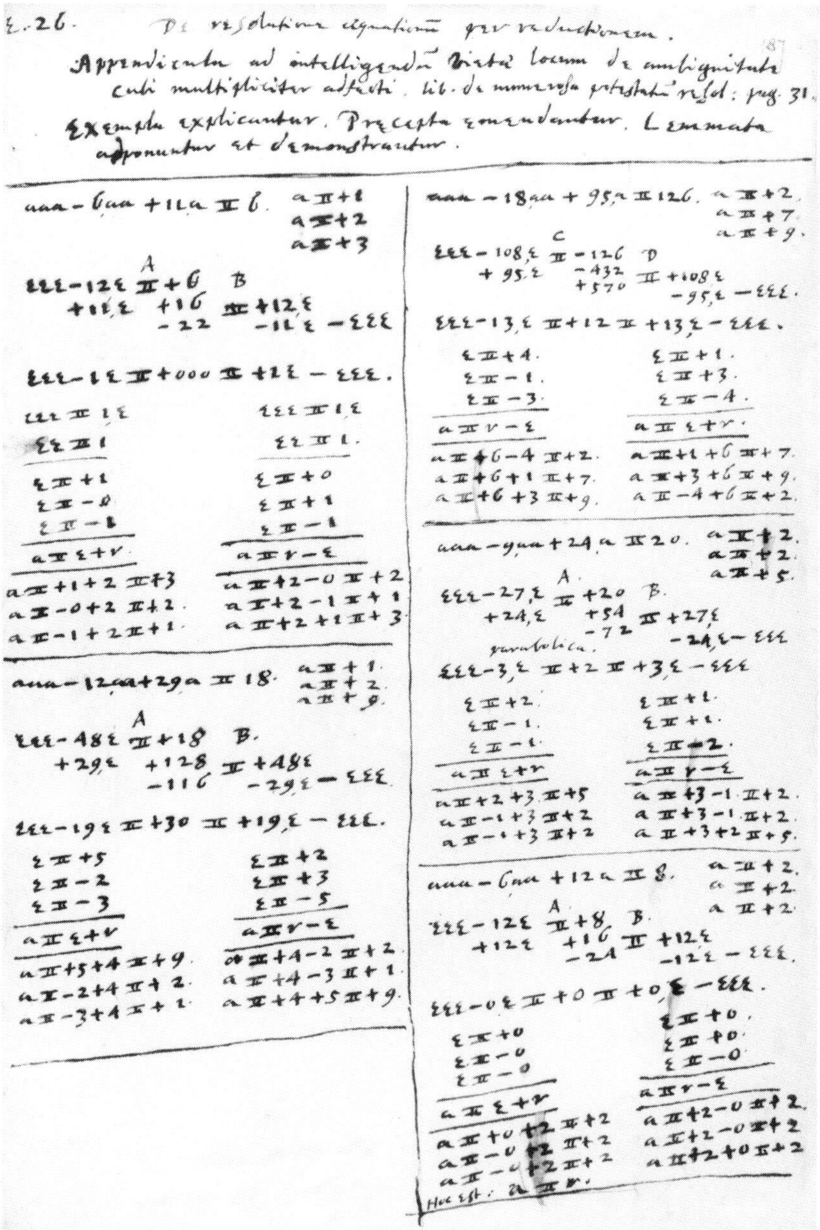

Fig. 12 Sheet *e*.26) of the *Treatise on equations*, Add MS 6783, f. 187.

On solving equations by reduction

⌣

Add MS 6783 f. 98

*e.*1)   *On solving equations by reduction*[1]

Preliminaries on the degrees, types and cases of the forthcoming equations[2]

1 $+$	31 $+++ +.+$
$-$	$+++ +.-$
3 $+.+$	$+++ -.+$
$+.-$	$+++ -.-$
$-.+$	$++ -+.+$
$-.-$	$++ -+.-$
7 $++.+$	$++ --.+$
$++.-$	$++ --.-$
$+-.+$	$+- ++.+$
$+-.-$	$+- ++.-$
$-+.+$	$+- +-.+$
$-+.-$	$+- +-.-$
$--.+$	$+- -+.+$
$--.-$	$+- -+.-$
	$+- --.+$
	$+- --.-$
	$-+ ++.+$
15 $+++.+$	$-+ ++.-$
$+++.-$	$-+ +-.+$
$++-.+$	$-+ +-.-$
$++-.-$	$-+ -+.+$
$+-+.+$	$-+ -+.-$
$+-+.-$	$-+ --.+$
$+--.+$	$-+ --.-$
$+--.-$	$-- ++.+$
$-++.+$	$-- ++.-$
$-++.-$	$-- +-.+$
$-+-.+$	$-- +-.-$
$-+-.-$	$-- -+.+$
$--+.+$	$-- -+.-$
$--+.-$	$-- --.+$
$---.+$	$-- --.-$
$---.-$	

1	1	1	1
2	1	2	1
	1	1 2	3
4	1	3	1
	2	1 3	3
		2 3	3
	1	1 2 3	7
8	1	4	1
	3	1 4	3
		2 4	3
		3 4	3
	3	1 2 4	7
		1 3 4	7
		2 3 4	7
	1	1 2 3 4	15
16	1	5	1
	4	1 5	3
		2 5	3
		3 5	3
		4 5	3
	6	1 2 5	7
		1 3 5	7
		1 4 5	7
		2 3 5	7
		2 4 5	7
		3 4 5	7
	4	1 2 3 5	15
		1 2 4 5	15
		1 3 4 5	15
		2 3 4 5	15
	1	1 2 3 4 5	31

32	1	6	1
	5	1 6	3
		2 6	3
		3 6	3
		4 6	3
		5 6	3
	10	1 2 6	7
		1 3 6	7
		1 4 6	7
		1 5 6	7
		2 3 6	7
		2 4 6	7
		2 5 6	7
		3 4 6	7
		3 5 6	7
		4 5 6	7
	10	1 2 3 6	15
		1 2 4 6	15
		1 2 5 6	15
		1 3 4 6	15
		1 3 5 6	15
		1 4 5 6	15
		2 3 4 6	15
		2 3 5 6	15
		2 4 5 6	15
		3 4 5 6	31
	5	1 2 3 4 6	31
		1 2 3 5 6	31
		1 2 4 5 6	31
		1 3 4 5 6	31
		2 3 4 5 6	31
	1	1 2 3 4 5 6	63

etc.[3]

A collection and summary of the enumeration:[4]

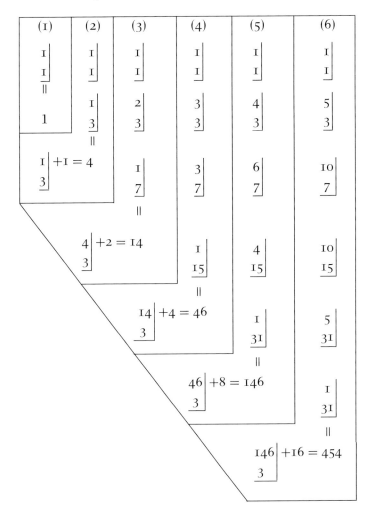

If equations with all negative coefficients are counted, the progression of the sums will be:

$1 + 1 = 2$

$4 + 2 = 6$

$14 + 4 = 18$

$46 + 8 = 54$

$146 + 16 = 162$

$454 + 32 = 486$ etc.

where the sums are in triple ratio.

e.2)   *On solving equations by reduction*

The ordering of the forthcoming equations by classes, types and cases

$x = a$

$xz = aa$

$xz = + ba + aa$

$xz = - ba + aa$

$xz = + ba - aa$

$xxz = aaa$

$xxz = + bca + aaa$

$xxz = - bca + aaa$

$xxz = + bca - aaa$

$xxz = + baa + aaa$

$xxz = - baa + aaa$

$xxz = + baa - aaa$

$xxz = + cda + baa + aaa$

$xxz = - cda + baa + aaa$

$xxz = + cda - baa + aaa$

$xxz = - cda - baa + aaa$

$xxz = + cda + baa - aaa$

$xxz = - cda + baa - aaa$

$xxz = + cda - baa - aaa$

$xxxz = aaaa$

$xxxz = + bbca + aaaa$

$xxxz = - bbca + aaaa$

$xxxz = + bbca - aaaa$

$xxxz = + bcaa + aaaa$

$xxxz = - bcaa + aaaa$

$xxxz = + bcaa - aaaa$

$xxxz = + baaa + aaaa$

$xxxz = - baaa + aaaa$

$xxxz = + baaa - aaaa$

$xxxz = + ddfa + bcaa + aaaa$

$xxxz = - ddfa + bcaa + aaaa$

$xxxz = + ddfa - bcaa + aaaa$

$xxxz = - ddfa - bcaa + aaaa$

$xxxz = + ddfa + bcaa - aaaa$

$xxxz = - ddfa + bcaa - aaaa$

$xxxz = + ddfa - bcaa - aaaa$

$xxxz = + ccda + baaa + aaaa$

$xxxz = - ccda + baaa + aaaa$

$xxxz = + ccda - baaa + aaaa$

$xxxz = - ccda - baaa + aaaa$

$xxxz = + ccda + baaa - aaaa$

$xxxz = - ccda + baaa - aaaa$

$xxxz = + ccda - baaa - aaaa$

$xxxz = + cdaa + baaa + aaaa$

$xxxz = - cdaa + baaa + aaaa$

$xxxz = + cdaa - baaa + aaaa$

$xxxz = - cdaa - baaa + aaaa$

$xxxz = + cdaa + baaa - aaaa$

$xxxz = - cdaa + baaa - aaaa$

$xxxz = + cdaa - baaa - aaaa$

*e.*3)   *On solving equations by reduction*

The ordering of the equations

$$xxxz = + \textit{ffga} + \textit{cdaa} + \textit{baaa} + \textit{aaaa}$$
$$xxxz = - \textit{ffga} + \textit{cdaa} + \textit{baaa} + \textit{aaaa}$$
$$xxxz = + \textit{ffga} - \textit{cdaa} + \textit{baaa} + \textit{aaaa}$$
$$xxxz = - \textit{ffga} - \textit{cdaa} + \textit{baaa} + \textit{aaaa}$$
$$xxxz = + \textit{ffga} + \textit{cdaa} - \textit{baaa} + \textit{aaaa}$$
$$xxxz = - \textit{ffga} + \textit{cdaa} - \textit{baaa} + \textit{aaaa}$$
$$xxxz = + \textit{ffga} - \textit{cdaa} - \textit{baaa} + \textit{aaaa}$$
$$xxxz = - \textit{ffga} - \textit{cdaa} - \textit{baaa} + \textit{aaaa}$$
$$xxxz = + \textit{ffga} + \textit{cdaa} + \textit{baaa} - \textit{aaaa}$$
$$xxxz = - \textit{ffga} + \textit{cdaa} + \textit{baaa} - \textit{aaaa}$$
$$xxxz = + \textit{ffga} - \textit{cdaa} + \textit{baaa} - \textit{aaaa}$$
$$xxxz = - \textit{ffga} - \textit{cdaa} + \textit{baaa} - \textit{aaaa}$$
$$xxxz = + \textit{ffga} + \textit{cdaa} - \textit{baaa} - \textit{aaaa}$$
$$xxxz = - \textit{ffga} + \textit{cdaa} - \textit{baaa} - \textit{aaaa}$$
$$xxxz = + \textit{ffga} - \textit{cdaa} - \textit{baaa} - \textit{aaaa}$$

$$xxxxz = \textit{aaaaa}$$

$$xxxxz = + \textit{bbbca} + \textit{aaaaa}$$
$$xxxxz = - \textit{bbbca} + \textit{aaaaa}$$
$$xxxxz = + \textit{bbbca} - \textit{aaaaa}$$
$$xxxxz = + \textit{bbcaa} + \textit{aaaaa}$$
$$xxxxz = - \textit{bbcaa} + \textit{aaaaa}$$
$$xxxxz = + \textit{bbcaa} - \textit{aaaaa}$$

$$xxxxz = + \textit{bcaaa} + \textit{aaaaa}$$
$$xxxxz = - \textit{bcaaa} + \textit{aaaaa}$$
$$xxxxz = + \textit{bcaaa} - \textit{aaaaa}$$

$$xxxxz = + \textit{baaaa} + \textit{aaaaa}$$
$$xxxxz = - \textit{baaaa} + \textit{aaaaa}$$
$$xxxxz = + \textit{baaaa} - \textit{aaaaa}$$

etc.

e.3.2)   *Examples of equations in numbers*[5]

$2 = + a$	$a = + 2$
$2 = - a$	$a = - 2$
$4 = + aa$	$aa = + 4 \qquad a = + 2, - 2$
$4 = - aa$	$aa = - 4$
$8 = + 2a + aa$	$a = + 2, - 4$
$8 = - 2a + aa$	$a = + 4, - 2$
$8 = - 6a - aa$	$a = - 2, - 4$
$8 = + 6a - aa$	$a = + 2, + 4$
$8 = + aaa$	$a = + 2$
$8 = - aaa$	$a = - 2$
$20 = + 6a + aaa$	$a = + 2$
$20 = - 6a - aaa$	$a = - 2$
$20 = - 21a + aaa$	$a = + 5, - 1, - 4$
$20 = + 21a - aaa$	$a = - 5, + 1, + 4$
$40 = - 6a + aaa$	$a = + 4$
$40 = + 6a - aaa$	$a = - 4$
$972 = + 21aa + aaa$	$a = + 6, - 9, - 18$
$972 = + 21aa - aaa$	$a = - 6, + 9, + 18$
$16 = - 3aa - aaa$	$a = - 4$
$16 = - 3aa + aaa$	$a = + 4$
$20 = + 3aa + aaa$	$a = + 2$
$20 = + 3aa - aaa$	$a = - 2$

e.3.3)   *Examples of equations in numbers*

$60 = + 8a + 9aa + aaa$	$a = + 2, - 5, - 6$
$60 = - 8a + 9aa - aaa$	$a = - 2, + 5, + 6$
$6 = - 11a - 6aa - aaa$	$a = - 1, - 2, - 3$
$6 = + 11a - 6aa + aaa$	$a = + 1, + 2, + 3$
$588 = + 8a + 9aa + aaa$	$a = + 6$
$588 = - 8a + 9aa - aaa$	$a = - 6$
$12 = - 11a - 6aa - aaa$	$a = - 4$
$12 = + 11a - 6aa + aaa$	$a = + 4$
$30 = - 1a + 6aa + aaa$	$a = + 2, - 3, - 5$
$30 = + 1a + 6aa - aaa$	$a = - 2, + 3, + 5$
$100 = + 60a - 3aa - aaa$	$a = - 10, + 2, + 5$
$100 = - 60a - 3aa + aaa$	$a = + 10, - 2, - 5$
$888 = - 1a + 6aa + aaa$	$a = + 8$
$888 = + 1a + 6aa - aaa$	$a = - 8$
$5600 = + 60a - 3aa - aaa$	$a = - 20$
$5600 = - 60a - 3aa + aaa$	$a = + 20$

e.3.4)   *Examples of equations in numbers*

$16 = + aaaa$ aa $= + 4, - 4$		$a = + 2, - 2$
$16 = - aaaa$ $aaaa = - 16$		
$186 = + 185a + aaaa$ $a = + 1, - 6$		
$186 = - 185a + aaaa$ $a = - 1, + 6$		
$14 = - 15a - aaaa$ $a = - 1, - 2$		
$14 = + 15a - aaaa$ $a = + 1, + 2$		
$36 = + 5aa + aaaa$ $aa = + 4, - 9,$	$a = + 2, - 2$	
$36 = - 5aa + aaaa$ $aa = - 4, + 9$	$a = - 3, + 3$	
$36 = - 13aa - aaaa$ $aa = - 4, - 9$		
$36 = + 13aa - aaaa$ $aa = + 4, + 9,$	$a = + 2, + 3, - 2, - 3$	
$9262 = + 20aaa + aaaa$	$a = + 7, - 21$	
$9262 = - 20aaa + aaaa$	$a = - 7, + 21$	
$59319 = - 40aaa - aaaa$	$a = - 13, - 39$	
$59319 = + 40aaa - aaaa$	$a = + 13, + 39$	
$384 = + 176a + 4aa + aaaa$	$a = + 2, - 6$	
$384 = - 176a + 4aa + aaaa$	$a = - 2, + 6$	
$150 = - 95a - 6aa - aaaa$	$a = - 2, - 3$	
$150 = + 95a - 6aa - aaaa$	$a = + 2, + 3$	
$36 = + 60a - 25aa + aaaa$	$a = + 1, + 2, + 3, - 6$	
$36 = - 60a - 25aa + aaaa$	$a = - 1, - 2, - 3, + 6$	
$105 = - 64a + 42aa - aaaa$	$a = + 3, + 5, - 1, - 7$	
$105 = + 64a + 42aa - aaaa$	$a = - 3, - 5, + 1, + 7$	
$1155 = + 136a - 6aa + aaaa$	$a = + 5, - 7$	
$1155 = - 136a - 6aa + aaaa$	$a = - 5, + 7$	
$29319 = - 8000a + 600aa - aaaa$	$a = - 3, - 29$	
$29319 = + 8000a + 600aa - aaaa$	$a = + 3, + 29$	

e.3.5) *Examples of equations in numbers*

$144 = +48a + 4aaa + aaaa$	$a = +2, -6$
$144 = -48a - 4aaa + aaaa$	$a = -2, +6$
$108 = -27a - 4aaa - aaaa$	$a = -3, -4$
$108 = +27a + 4aaa - aaaa$	$a = +3, +4$
$81 = +90a - 10aaa + aaaa$	$a = +1, +3, +9, -3$
$81 = -90a + 10aaa + aaaa$	$a = -1, -3, -9, +3$
$80 = -92a + 11aaa - aaaa$	$a = +4, +10, -1, -2$
$80 = +92a - 11aaa - aaaa$	$a = -4, -10, +1, +2$
$45 = +48a - 4aaa + aaaa$	$a = +1, -3$
$45 = -48a + 4aaa + aaaa$	$a = -1, +3$
$87 = -92a + 4aaa - aaaa$	$a = -1, -3$
$87 = +92a - 4aaa - aaaa$	$a = +1, +3$
$576 = +100aa + 20aaa + aaaa$	$a = +2, -4, -6, -12$
$576 = +100aa - 20aaa + aaaa$	$a = -2, +4, +6, +12$
$288 = -22aa - 14aaa - aaaa$	$a = -4, -12$
$288 = -22aa + 14aaa - aaaa$	$a = +4, +12$
$512 = -8aa + 6aaa + aaaa$	$a = +4, -8$
$512 = -8aa - 6aaa + aaaa$	$a = -4, +8$
$288 = +62aa - 7aaa - aaaa$	$a = +3, +4, -2, -12$
$288 = +62aa + 7aaa - aaaa$	$a = -3, -4, +2, +12$

e.3.6) *Examples of equations in numbers*

$48 = +2a + 33aa + 12aaa + aaaa$	$a = +1, -2, -3, -8$
$48 = -2a + 33aa - 12aaa + aaaa$	$a = -1, +2, +3, +8$
$360 = -526a - 189aa - 24aaa - aaaa$	$a = -1, -4, -9, -10$
$360 = +526a - 189aa + 24aaa - aaaa$	$a = +1, +4, +9, +10$
$192 = +86a + 89aa + 16aaa + aaaa$	$a = +1, -3$
$192 = -86a + 89aa - 16aaa + aaaa$	$a = -1, +3$
$44 = -67a - 30aa - 8aaa - aaaa$	$a = -1, -4$
$44 = +67a - 30aa + 8aaa - aaaa$	$a = +1, +4$

$$300 = -40a + 59aa + 16aaa + aaaa \qquad a = +2, -3, -5, -10$$
$$300 = +40a + 59aa - 16aaa + aaaa \qquad a = -2, +3, +5, +10$$
$$100 = +120a - 7aa - 12aaa - aaaa \qquad a = +1, +2, -5, -10$$
$$100 = -120a - 7aa + 12aaa - aaaa \qquad a = -1, -2, +5, +10$$

$$480 = -10a + 89aa + 16aaa + aaaa \qquad a = +2, -3$$
$$480 = +10a + 89aa - 16aaa + aaaa \qquad a = -2, +3$$
$$200 = +90a - 9aa - 12aaa - aaaa \qquad a = -5, -10$$
$$200 = -90a - 9aa + 12aaa - aaaa \qquad a = +5, +10$$
$$200 = +270a - 57aa - 12aaa - aaaa \qquad a = +1, +2$$
$$200 = -270a - 57aa + 12aaa - aaaa \qquad a = -1, -2$$

$$900 = -540a - 77aa + 4aaa + aaaa \qquad a = +10, -3, -5, -6$$
$$900 = +540a - 77aa - 4aaa + aaaa \qquad a = -10, +3, +5, +6$$
$$105 = +104a + 10aa - 8aaa - aaaa \qquad a = +1, +3, -5, -7$$
$$105 = -104a + 10aa + 8aaa - aaaa \qquad a = -1, -3, +5, +7$$

$$1800 = -750a - 47aa + 4aaa + aaaa \qquad a = +10, -3$$
$$1800 = +750a - 47aa - 4aaa + aaaa \qquad a = -10, +3$$
$$175 = +80a + 8aa - 8aaa - aaaa \qquad a = -5, -7$$
$$175 = -80a + 8aa + 8aaa - aaaa \qquad a = +5, +7$$
$$81 = +84a + 2aa - 4aaa - aaaa \qquad a = +1, +3$$
$$81 = -84a + 2aa + 4aaa - aaaa \qquad a = -1, -3$$

$$132 = +236a - 121aa + 16aaa + aaaa \qquad a = +1, +2, +3, -22$$
$$132 = -236a - 121aa - 16aaa + aaaa \qquad a = -1, -2, -3, +22$$
$$225 = -120a + 98aa - 8aaa - aaaa \qquad a = +3, +5, -1, -15$$
$$225 = +120a + 98aa + 8aaa - aaaa \qquad a = -3, -5, +1, +15$$

$$330 = +425a - 112aa + 16aaa + aaaa \qquad a = +1, -22$$
$$330 = -425a - 112aa - 16aaa + aaaa \qquad a = -1, +22$$
$$270 = -168a + 95aa - 8aaa - aaaa \qquad a = -1, -15$$
$$270 = +168a + 95a + 8aaa - aaaa \qquad a = +1, +15$$

The sum of all: 120

*e.*4)   *On solving equations by reduction*

Two other canonical equations besides those which have been treated above (in sheets *d.*)

The two canonical forms are: I. $rrr - qqq = + 3qra + aaa$ $a = r - q$

                    2. $rrr + qqq = - 3qra + aaa$ $a = r + q$

The generation of the first:

Let $r - a = q$

then $r - q = a$

and $-r + a = -q$

therefore $\begin{vmatrix} r - a \\ r - a \\ r - a \end{vmatrix} = rrr - 3rra + 3raa - aaa = qqq$

$-3rra + 3raa = \begin{vmatrix} -r \\ 3ra \end{vmatrix} \begin{vmatrix} +a \\ 3ra \end{vmatrix} = \begin{vmatrix} -q \\ 3ra \end{vmatrix} = -3qra$

therefore $rrr - qqq = +3qra + aaa$ $a = r - q$

Let $a = r - q$

then $rrr - qqq = \begin{vmatrix} + 3qr \\ r - q \end{vmatrix} \begin{vmatrix} + r - q \\ r - q \\ r - q \end{vmatrix} = rrr - 3rrq + 3rqq - qqq$

                                  $+3rrq - 3rqq$

and it is so. Therefore the proposition is true.

Alternatively, let $r + a = q$

then $a = q - r$

therefore $\begin{vmatrix} r + a \\ r + a \\ r + a \end{vmatrix} = rrr + 3rra + 3raa + aaa = qqq$

$+3rra + 3raa = \begin{vmatrix} +r \\ 3ra \end{vmatrix} \begin{vmatrix} +a \\ 3ra \end{vmatrix} = \begin{vmatrix} +q \\ 3ra \end{vmatrix} = +3qra$

therefore $3qra + aaa = qqq - rrr$

or $qqq - rrr = +3qra + aaa$

It may be demonstrated as above that $a = q - r$ [satisfies this equation].

The generation of the second:

Let $\quad a - r = q$

then $\quad a = q + r$

and $\quad - a + r = - q$

therefore $\quad \begin{vmatrix} a - r \\ a - r \\ a - r \end{vmatrix} = aaa - 3raa + 3rra - rrr = qqa$

$$- 3raa + 3rra = \begin{vmatrix} -a \\ 3ra \end{vmatrix} + r \begin{vmatrix} \, \\ 3ra \end{vmatrix} = \begin{vmatrix} -q \\ 3ra \end{vmatrix} = -3qra$$

therefore $\quad qqq + rrr = -3qra + aaa \qquad\qquad\qquad a = r + q$

Let $\quad a = r + q$

then $\quad qqq + rrr = \begin{vmatrix} -3qr \\ r+q \end{vmatrix} + \begin{vmatrix} r+q \\ r+q \\ r+q \end{vmatrix} = \begin{matrix} rrr + 3rrq + 3rqq + qqq \\ - 3rrq - 3rqq \end{matrix}$

and it is so. Therefore the proposition is true.

*e.*4)   *On solving equations by reduction*

The equation to be solved:	$ggh = + dfa + aaa$	
or	$2ccc = + 3bba + aaa$	
The canon for finding the roots is:	$qqq - rrr = + 3qra + aaa$	$a = q - r$

Let $e = q$ so $\dfrac{bb}{e} = r$

therefore $eee - \dfrac{bbbbbb}{eee} = 2ccc$

$eeeee - bbbbbb = 2ccceee$

$eeeee - 2ccceee = bbbbbb$

     [add] $ccccc$ [to both sides]

$eeeee - 2ccceee + ccccc = bbbbbb + ccccc$

1st $eee - ccc = \sqrt{(bbbbbb + ccccc)}$

  $eee = \sqrt{(bbbbbb + ccccc)} + ccc = qqq$

  $e = \sqrt[3]{\sqrt{bbbbbb + ccccc} + ccc} = q$

2nd $ccc - eee = \sqrt{(bbbbbb + ccccc)}$

so $ccc - \sqrt{(bbbbbb + ccccc)} = eee$

But ccc is not greater than $\sqrt{(bbbbbb + ccccc)}$, so there is no second root unless it is negative. Then it will be $- rrr$ as will become apparent below.

Let $e = r$ so $\dfrac{bb}{e} = q$

therefore $\dfrac{bbbbbb - eee}{eee} = 2ccc$

and $bbbbbb - eeeee = 2ccceee$

so $bbbbbb = 2ccceee + eeeee$

  [add] $ccccc$ [to both sides]

then $bbbbbb + ccccc = ccccc + 2ccceee + eeeee$

$\sqrt{(bbbbbb + ccccc)} = ccc + eee$

$\sqrt{(bbbbbb + ccccc)} - ccc = eee = rrr$

$\sqrt[3]{\sqrt{bbbbbb + ccccc} - ccc} = e = r$

Here it appears that $- rrr$ is equal to the second root above. And all the solutions may be formed from these two roots.

But $a = q - r$.

Therefore the solution formula[6] found will be:

$$a = \sqrt[3]{\sqrt{bbbbbb + ccccc} + ccc} - \sqrt[3]{\sqrt{bbbbbb + ccccc} - ccc}$$

or $\left(\text{since } \dfrac{bb}{q} = r\right)$

$$a = \sqrt[3]{\sqrt{bbbbbb + ccccc} + ccc} - \dfrac{bb}{\sqrt[3]{\sqrt{bbbbbb + ccccc} + ccc}}$$

$$a = \sqrt[3]{\sqrt{\dfrac{dfdfdf}{27} + \dfrac{gghggh}{4}} + \dfrac{ggh}{2}} - \sqrt[3]{\sqrt{\dfrac{dfdfdf}{27} + \dfrac{gghggh}{4}} - \dfrac{ggh}{2}}$$

$20 = + 6a + aaa$ $a = \sqrt[3]{\sqrt{108} + 10} - \sqrt[3]{\sqrt{108} - 10}$
$\qquad\qquad\qquad = (\sqrt{3} + 1) - (\sqrt{3} - 1)$
$\qquad\qquad\qquad = \sqrt{3} + 1 - \sqrt{3} + 1$
$\qquad\qquad\qquad = 2$

$26 = + 9a + aaa$ $a = \sqrt[3]{14 + 13} - \sqrt[3]{14 - 13} = 2$

$7 = + 6a + aaa$ $a = \sqrt[3]{\dfrac{9}{2} + \dfrac{7}{2}} - \sqrt[3]{\dfrac{9}{2} - \dfrac{7}{2}} = 1$

e.6)   *On solving equations by reduction*

The equation to be solved: $ggh = -dfa + aaa$

or $2ccc = -3bba + aaa$

The canon for finding the roots is: $qqq + rrr = -3qra + aaa$ $a = q + r$

Let $e = q$ so $\dfrac{bb}{e} = r$

then $eee + \dfrac{bbbbbb}{eee} = 2ccc$

and $eeeee + bbbbbb = 2ccceee$

so $eeeee - 2ccceee = -bbbbbb$

         [add] $cccccc$ [to both sides]

then $eeeee - 2ccceee + cccccc = +cccccc - bbbbbb$

1st $eee - ccc = \sqrt{(cccccc - bbbbbb)}$

$eee = ccc + \sqrt{(cccccc - bbbbbb)}$

$e = \sqrt[3]{ccc + \sqrt{cccccc - bbbbbb}} = q$

2nd $ccc - eee = \sqrt{(cccccc - bbbbbb)}$

$ccc - \sqrt{(cccccc - bbbbbb)} = eee$

$e = \sqrt[3]{ccc + \sqrt{cccccc - bbbbbb}}$

The sum of these two quadratic roots is $2ccc$. One is qqq, the other rrr. Whence the solution may be found, from these two roots; and it may be done thus as above.

Let $e = r$ so $\dfrac{bb}{e} = q$

then $\dfrac{bbbbbb}{eee} + eee = 2ccc$

and $bbbbbb + eeeee = 2ccceee$

so $eeeee - 2ccceee = -bbbbbb$

and from here, all will be as above.

Therefore if the first root is q, the second will be r.

Or one root is q, and the other is r.

But $a = q + r$.

Therefore the solution formula found will be:

$$a = \sqrt[3]{ccc + \sqrt{ccccc - bbbbbb}} + \sqrt[3]{ccc - \sqrt{ccccc - bbbbbb}}$$

or $\left(\text{since } \dfrac{bb}{q} = r \right)$

$$a = \sqrt[3]{ccc + \sqrt{ccccc - bbbbbb}} + \dfrac{bb}{\sqrt[3]{ccc + \sqrt{ccccc - bbbbbb}}}$$

$$a = \sqrt[3]{\dfrac{ggh}{2} + \sqrt{\dfrac{gghggh}{4} - \dfrac{dfdfdf}{27}}} + \sqrt[3]{\dfrac{ggh}{2} - \sqrt{\dfrac{gghggh}{4} - \dfrac{dfdfdf}{27}}}$$

$40 = -6a + aaa$ $\quad a = \sqrt[3]{20 + \sqrt{392}} + \sqrt[3]{20 - \sqrt{392}}$
$$ = 2 + \sqrt{2} + 2 - \sqrt{2}$$
$$ = 4$$

$72 = -24a + aaa$ $\quad a = \sqrt[3]{36 + 28} + \sqrt[3]{36 - 28} = 6$

$9 = -6a + aaa$ $\quad a = \sqrt[3]{\dfrac{9}{2} + \dfrac{7}{2}} + \sqrt[3]{\dfrac{9}{2} - \dfrac{7}{2}} = 3$

e.7)   *On solving equations by reduction*

Another way

The equation to be solved: $2ccc = -3bba + aaa$

The canon: $qqq + rrr = -3qra + aaa$ $a = q + r$

From the given equation, we must seek $q + r = a$.

First let r be sought.

$qr = bb$

therefore $r = \dfrac{bb}{q}$

$qqq + rrr = 2ccc$

$rrr = 2ccc - qqq$

therefore $r = \sqrt[3]{2ccc - qqq}$

therefore $\dfrac{bb}{q} = \sqrt[3]{2ccc - qqq}$

$\dfrac{bbbbbb}{qqq} = 2ccc - qqq$

$bbbbbb = 2cccqqq - qqqqqq$

$qqqqqq - 2cccqqq = -bbbbbb$

         [add] $cccccc$ [to both sides]

$qqqqqq - 2cccqqq + cccccc = cccccc - bbbbbb$

$qqq - ccc = \sqrt{(cccccc - bbbbbb)}$

$qqq = ccc + \sqrt{(cccccc - bbbbbb)}$

therefore 1st $q = \sqrt[3]{ccc + \sqrt{cccccc - bbbbbb}}$

           2nd $q = \sqrt[3]{ccc - \sqrt{cccccc - bbbbbb}}$

But $\quad r = \sqrt[3]{2ccc - qqq}$

therefore, from 1st q, $\quad r = \sqrt[3]{2ccc - ccc - \sqrt{cccccc - bbbbbb}}$

that is $\quad r = \sqrt[3]{ccc - \sqrt{cccccc - bbbbbb}}$, the same as 2nd q

therefore $\quad q + r = \sqrt[3]{ccc + \sqrt{cccccc - bbbbbb}} + \sqrt[3]{ccc - \sqrt{cccccc - bbbbbb}}$

From 2nd q, $\quad r = \sqrt[3]{2ccc - ccc + \sqrt{cccccc - bbbbbb}}$

that is $\quad r = \sqrt[3]{ccc + \sqrt{cccccc - bbbbbb}}$

therefore $\quad r + q = \sqrt[3]{ccc + \sqrt{cccccc - bbbbbb}} + \sqrt[3]{ccc - \sqrt{cccccc - bbbbbb}}$

But also, $r = \dfrac{bb}{q}$

therefore from 1st q:

$$q + r = \sqrt[3]{ccc + \sqrt{cccccc - bbbbbb}} + \frac{bb}{\sqrt[3]{ccc + \sqrt{cccccc - bbbbbb}}}$$

and from 2nd q:

$$q + r = \sqrt[3]{ccc - \sqrt{cccccc - bbbbbb}} + \frac{bb}{\sqrt[3]{ccc - \sqrt{cccccc - bbbbbb}}}$$

But $\quad q + r = a$.

Therefore the solution formula can be found, and is the same as in the sheet above [e.6)].

e.7.2)   *On solving equations by reduction*

Another way

The equation to be solved:	$2ccc = -3bba + aaa$
The canon:	$qqq + rrr = -3qra + aaa$ $a = q + r$

Let $e = q$ so $eee = qqq$

therefore $2ccc - eee = rrr$

and $qqq + rrr = 2ccc$

therefore $\dfrac{\sqrt[3]{2ccc - eee}}{\sqrt[3]{eee}} = qr = bb$

therefore $\sqrt[3]{2ccceee - eeeeee} = bb$

$2ccceee - eeeeee = bbbbbb$

$-2ccceee + eeeeee = -bbbbbb$

[add] $cccccc$ [to both sides]

$cccccc - 2ccceee + eeeeee = cccccc - bbbbbb$

$1^{st})$ $ccc - eee = \sqrt{(cccccc - bbbbbb)}$

$2^{nd})$ $- ccc + eee = \sqrt{(cccccc - bbbbbb)}$

therefore $1^{st})$ $ccc - \sqrt{(cccccc - bbbbbb)} = eee$

$2^{nd})$ $eee = ccc - \sqrt{(cccccc - bbbbbb)}$

The sum [of these two roots] is $2ccc$.

Therefore one root is qqq, and the other is rrr.

But $q + r = a$.

Therefore the solution formula found will be (as above):

$$a = \sqrt[3]{ccc + \sqrt{cccccc - bbbbbb}} + \sqrt[3]{ccc - \sqrt{cccccc - bbbbbb}}$$

This must also be noted in passing, that $bbbbbb$ is rightly considered the cube of a square, not however the cube of a cube. Similarly $cccccc$ must be considered the square of a cube, and not the cube of a cube.[7]

For these expressions in letters are more useful than the corresponding forms obtained from the generation, which was done otherwise before, in canonicals instead; other expressions, however, will have their places, as there is opportunity.

e.8)   *On solving equations by reduction*

There are clearly three cases of the foregoing cubic equation:

$$2ccc = -3bba + aaa$$

In the first case we have:	$ccccc > bbbbb$	or	$c > b$
second:	$ccccc < bbbbb$	or	$c < b$
third:	$ccccc = bbbbb$	or	$c = b$

The first case may be called hyperbolic; the second case elliptic; the third case parabolic.

Corresponding to these three cases, there are three canonical forms:

Hyperbolic canon:	$qqq + rrr = -3qra + aaa$	$a = q + r$
Elliptic:	$qqr + qrr = -qqa - qra - rra + aaa$	$a = q + r$
Parabolic:	$2qqq = -3qqa + aaa$	$a = 2q$

The hyperbolic canon is generated above in (e.4), and the elliptic in (d.4).
The parabolic may be found from the hyperbolic or elliptic by changing r to q.

Then it must be demonstrated that these three canonical forms have the properties of the three above cases.
But first must come these two lemmas:

1. If three unequal quantities are in continued proportion, the sum of the two extremes is greater than twice the middle quantity.
2. If four unequal quantities are in continued proportion, the sum of the two extremes is greater than the sum of the two intermediate quantities.

Lemma 1.[8]

Suppose b, a, aa are in continued proportion, and suppose $b > a$.

$$\overline{b}$$

I say that $\quad b + \dfrac{aa}{b} > 2a$

that is $\quad bb + aa > 2ab$

so $\quad\quad bb - ba > ba - aa$

that is $\quad \dfrac{b - a}{b} > \dfrac{b - a}{a}$

so $\quad\quad b > a$

and it is so. Therefore the lemma is true.

Lemma 2 may be demonstrated in the same way.

Suppose b, c, d, f are in continued proportion.

I say that $b + f > c + d$

$b + d > 2c$ [by lemma 1]

$c + f > 2d$ [by lemma 1]

therefore $\quad b + c + d + f > 2c + 2d$

therefore $\quad b + f > c + d$

which was to be demonstrated.

Then, first, I say that:

$$\left.\begin{array}{c} \dfrac{qqq + rrr}{2} \\ qqq + rrr \\ \hline 2 \end{array}\right| \quad \begin{array}{c} > qr \\ qr \\ qr \end{array} \qquad \text{(if } r = q \text{ there will be equality)}$$

that is
$$\left.\begin{array}{c} qqq + rrr \\ \dfrac{qqq + rrr}{4} \end{array}\right| \quad \begin{array}{c} > \dfrac{4qrqrqr}{4} \end{array}$$

so $\quad \dfrac{qqqqqq + 2qqqrrr + rrrrrr}{4} > \dfrac{4qqqrrr}{4}$

so $\quad qqqqqq + rrrrrr > 2qqqrrr$

and it is so by the second lemma. Therefore the proposition is true.

Add MS 6783 f. 107

e.9)   *On solving equations by reduction*

Then, second, I say that:

$$\left.\begin{array}{c} \dfrac{qqr + qrr}{2} \\ qqr + qrr \\ \hline 2 \end{array}\right| \quad \begin{array}{c} < \\ \end{array} \left.\begin{array}{c} \dfrac{qq + qr + rr}{3} \\ \dfrac{qq + qr + rr}{3} \\ \dfrac{qq + qr + rr}{3} \end{array}\right| \qquad \text{(if } r = q \text{ there will be equality)}$$

that is
$$\left.\begin{array}{c} qqr + qrr \\ \dfrac{qqr + qrr}{4} \end{array}\right| \quad < \quad \left.\begin{array}{c} qq + qr + rr \\ qq + qr + rr \\ \dfrac{qq + qr + rr}{27} \end{array}\right|$$

that is $\quad \dfrac{qqqqrr + 2qqqrrr + qqrrrr}{4}$

$$< \dfrac{qqqqqq + 3qqqqqr + 6qqqqqr + 7qqqrrr + 6qqrrrr + 3qrrrrr + rrrrrr}{27}$$

so $27qqqqrr + 54qqqrrr + 27qqrrrrr$
  $< 4qqqqqqq + 12qqqqqqr + 24qqqqqrr + 28qqqrrr + 24qqrrrrr + 12qrrrrrr + rrrrrrr$

so $3qqqqqrr + 26qqqrrr + 3qqrrrrr < 4qqqqqqq + 12qqqqqqr + 12qrrrrrr + 4rrrrrrr$

that is $3qqqqqrr + 12qqqqrrr + 12qqqqrrr + 3qqrrrrr$
  $+ 1qqqrrr \quad + 1qqqrrr$
  $< 3qqqqqqq + 12qqqqqqr + 12qrrrrrr + 3rrrrrrr$
  $+1qqqqqqq \qquad\qquad + 1rrrrrrr$

that is $(3qqqqqrr + 3qqrrrrr) + (12qqqqrrr + 12qqqqrrr) + (qqqrrr + qqqrrr)$
  $< (3qqqqqqq + 3rrrrrrr) + (12qqqqqqr + 12qrrrrrr) + (qqqqqqq + rrrrrrr)$

and so it is by the lemma. Therefore the proposition is true.

Then, third, I say that: $\begin{array}{c|c} qqq & = qq \\ \underline{qqq} & qq \\ & \underline{qq} \end{array}$

that is, $qqq.qqq = qq.qq.qq$
and it is so. Therefore the proposition is true.

---

e.10)   *On solving equations by reduction*

The equation to be solved: $2ccc = -3bba + aaa$

The canon: $qqr = -qqa$

$+ qrr \quad - qra$

$- rra + aaa$ $a = q - r$

From the given equation, we must seek $q + r = a$.

First let r be sought.

$qq + qr + rr = 3bb$

$qr + rr = 3bb - qq$

   [add] $\dfrac{qq}{4}$ [to each side]

$\dfrac{qq}{4} + qr + rr = 3bb - qq + \dfrac{qq}{4}$

$\dfrac{q}{2} + r = \sqrt{3bb - \dfrac{3qq}{4}}$

therefore $r = \sqrt{3bb - \dfrac{3qq}{4}} - \dfrac{q}{2}$

$qqr + qrr = 2ccc$

$qr + rr = \dfrac{2ccc}{q}$

  [add] $\dfrac{qq}{4}$ [to each side]

$\dfrac{qq}{4} + qr + rr = \dfrac{2ccc}{q} + \dfrac{qq}{4}$

$\dfrac{q}{2} + r = \sqrt{\dfrac{2ccc}{q} + \dfrac{qq}{4}}$

therefore $r = \sqrt{\dfrac{2ccc}{q} + \dfrac{qq}{4}} - \dfrac{q}{2}$

Therefore $3bb - \dfrac{3qq}{4} = \dfrac{2ccc}{q} + \dfrac{qq}{4}$

$3bb - qq = \dfrac{2ccc}{q}$

therefore $3bbq - qqq = 2ccc.$

And if q is sought before r we will have

$3bbr - rrr = 2ccc$

and either equation [for q or r] is a conjugate of the given equation.

Another way, by a different canon, from $(d.4)$:

$qqr - qrr = -qqa + qra - rra + aaa$

yielding $a = q$.

---

a must be sought from the given equation.

First let q be sought.

$qr - qr + rr = 3bb$

$-qr + rr = 3bb - qq$

$\qquad$ [add] $\dfrac{qq}{4}$ [to each side]

$\dfrac{qq}{4} - qr + rr = 3bb - qq + \dfrac{qq}{4}$

$\dfrac{q}{2} - r = \sqrt{\dfrac{3bb - 3qq}{4}}$

There is no need to proceed further for r.

$qqr - qrr = 2ccc$

$qr - rr = \dfrac{2ccc}{q}$

$-qr + rr = -\dfrac{2ccc}{q}$

$\qquad$ [add] $\dfrac{qq}{4}$ [to each side]

$\dfrac{qq}{4} - qr + rr = \dfrac{qq}{4} - \dfrac{2ccc}{q}$

$\dfrac{q}{2} - r = \sqrt{\dfrac{qq}{4} - \dfrac{2ccc}{q}}$

Therefore $\quad \dfrac{3bb - 3qq}{4} = \dfrac{qq}{4} - \dfrac{2ccc}{q}$

$3bb - qq = -\dfrac{2ccc}{q}$

$3bbq - qqq = -2ccc$

therefore $\quad 2ccc = -3bbq + qqq$

the same equation as the one given.

Note.

It is clear therefore that the elliptic equation, the second case, cannot be solved by any canonical form other than the conjugate, or in the general way.[9] Also, the conjugate, as will be apparent after this, cannot be solved except in the same general way. Which has been sufficiently explained in its place in what has gone before. Nevertheless, this must be noted, if the elliptic equation can be solved by its conjugate, the sum of the two roots of the conjugate is the sought root of the elliptic equation.[10]

Add MS 6783 f. 109

*e.*11)   *On solving equations by reduction*

The equation to be solved: $2ccc = -3bba + aaa$

The canon: $2qqq = -3qqa + aaa$ $a = 2q$

In this third case we may put $c = b$

whence the form of the equation will be as the canonical equation:

$$2bbb = -3bba + aaa \qquad a = 2b$$

Therefore the solution formula is immediately obvious, and appears from this.

$16 = -12a + aaa$

$a = \sqrt[3]{8+0} + \sqrt[3]{8+0} = 4$ as in the hyperbolic form

or $a = 2\sqrt[3]{8} = 4$

or $a = \dfrac{16}{12/3} = \dfrac{16}{4} = 4$

or $a = \dfrac{16/12}{3} = \dfrac{48}{12} = 4$

or $a = 2\sqrt{\dfrac{12}{3}} = 4$

or otherwise, but there is no need.

Up to now we have dealt with the three cases of the equation:

$ggh = -dfa + aaa$

or $2ccc = -3bba + aaa$

e.12)   *On solving equations by reduction*

The equation to be solved:	$ggh = + dfa - aaa$	
or	$2ccc = + 3bba - aaa$	
The canon, from (*d*.3) is:	$qqr = + qqa$	
	$+ qrr \quad + qra$	$a = q$
	$+ rra - aaa$	$a = r$

From the given equation, there must be sought q and r.

First let r be sought.
$$qq + qr + rr = 3bb$$
$$qr + rr = 3bb - qq$$
etc. as (*e*.10).

Therefore	$3bbq - qqq = 2ccc$
and	$3bbr - rrr = 2ccc$

Or	$dfq - qqq = ggh$
	$dfr - rrr = ggh$

Either equation is the same as the one given.

Whence it is apparent that the given equation necessarily has two roots, as does the canon, that is $a = q$ and $a = r$.

Another way, by a different canon, from (*d*.4)

$qqr = + qqa$	
$- qrr \quad - qra$	$a = q - r$
$+ rra - aaa$	$a = r$

From the given equation there must be sought $q - r$ and r.

First let r be sought.
$$qq - qr + rr = 3bb$$
etc as (*e*.10).

Therefore	$2ccc = - 3bbq + qqq$
or	$ggh = - dfq + qqq$

This is the conjugate to the given equation, whence q is determined, and is equal to the sum of the roots of the given equation, $q - r$ and r.

Note on the solution:

Here it is apparent that the given equation can not be solved by canons other than its conjugate. Or in the general way etc. as was said above.

e.13)   *On solving equations by reduction*

The solution of the preceding equation $ggh = dfa - aaa$
given the solution of the conjugate.

The conjugate is $aaa - dfa = ggh$.
Let $a = x = q + r$ in the first canonical form.

I. The solution formula may be found from the first canon for the given
equation.
Let one of the roots sought be $e = q$
then the other will be $x - e = r$.

Because $qq + qr + rr = df$
therefore $ee + xe - ee + xx - 2xe + ee = df$
that is $xx - xe + ee = df$
but $xx > df$ since $\begin{array}{|c} q + r \\ q + r \end{array}$ $> qq + qr + rr$

therefore $xx - df = xe - ee$
therefore $ee - xe = df - xx$
$$[\text{add}]\ \frac{xx}{4}\ [\text{to each side}]$$

$$ee - xe + \frac{xx}{4} = df + \frac{xx}{4} - xx$$

$$1^{\text{st}}\quad e - \frac{x}{2} = \sqrt{df - \frac{3xx}{4}}$$

$$2^{\text{nd}}\quad -e + \frac{x}{2} = \sqrt{df - \frac{3xx}{4}}$$

Therefore $$1^{\text{st}}\quad e = \frac{x}{2} + \sqrt{df - \frac{3xx}{4}}$$

$$2^{\text{nd}}\quad e = \frac{x}{2} - \sqrt{df - \frac{3xx}{4}}$$

The sum of these two roots is x.
One is e, the other is $x - e$.
Therefore these two roots are the sought solution formulae.
(The product of the two roots is $xx - df$.)

Another way by the first canon.

Let one root be $e = q$

then the other will be $x - e = r$.

Because $qqq + qrr = ggh$

or
$$\begin{array}{c|c} q+r & x \\ \hline qr & xe-ee \end{array} = $$

therefore $\quad xxe - xee = ggh$

$$xe - ee = \frac{ggh}{x}$$

$$- xe + ee = - \frac{ggh}{x}$$

$$[add] \ \frac{xx}{4} \ [to \ each \ side]$$

$$\frac{xx}{4} - xe + ee = \frac{xx}{4} - \frac{ggh}{x}$$

$$1^{st} \quad x - \frac{e}{2} = \sqrt{\frac{xx}{4} - \frac{ggh}{x}}$$

$$2^{nd} \quad - x + \frac{e}{2} = \sqrt{\frac{xx}{4} - \frac{ggh}{x}}$$

$$1^{st} \quad e = x - \sqrt{\frac{xx}{4} - \frac{ggh}{x}}$$

$$2^{nd} \quad e = x + \sqrt{\frac{xx}{4} - \frac{ggh}{x}}$$

The sum of these two roots is x.

One is e, the other is $x - e$.

Therefore these two roots are also solution formulae.

But the previous formulae are more useful for solving the equation.

$$aaa - 63a = 162 = 63a - aaa$$

$$\begin{array}{ccc} a = 9 & & a = 3 \\ & & a = 6 \end{array}$$

*e.*14)   *On solving equations by reduction*

II. By the second canon for the given equation.

In the second canon x will be q.
Then if one root sought is supposed $e = r$
the other will be $x - e = q - r$.
Then $xx - xe + ee = df$
etc. as in the first method.

III. Without the canon, by the rule in $(d.19)$ as Analysts did before us.
The conjugate equation is: $aaa - dfa = ggh$.
Let $a = x$
therefore $xxx - dfx = ggh$
and $xxx - ggh = dfx$.

The given equation is $dfa - aaa = ggh$
therefore $dfa - ggh = aaa$
            add xxx [to each side]
therefore $dfa + xxx - ggh = aaa + xxx$

but $xxx - ggh = dfx$ [see above][11]
therefore $dfa + dfx = aaa + xxx$
therefore $\dfrac{dfa + dfx}{a + x} = \dfrac{aaa + xxx}{a + x}$

therefore $df = xx - xa + aa$ by the rule in $(d.19)$
etc. as in the first method.

Up to now we have treated the first three binomial cubic equations[12] and their solutions:

$ggh = + dfa + aaa$
$ggh = - dfa + aaa$
$ggh = + dfa - aaa$

Omitted from these derivations are linear and quadratic equations, as being sufficiently known.

It is to be expected in what follows that there is some method of reduction to one or other of the preceding forms, whence the solution may be found.

---

e.15)   *On solving equations by reduction*

The equation to be reduced: $ccd = -baa + aaa$

or $2xxx = -3raa + aaa$

I. $aaa - 3raa = 2xxx$

add $+3rra - rrr$ [to each side]

therefore $aaa - 3raa + 3rra - rrr = +2xxx + 3rra$
$$- rrr$$

Let $a - r = e$

so $a = e + r$

$3rra = +3rre + 3rrr$

therefore $eee - 3rre = +2xxx$ (A)
$$+ 2rrr, \quad \text{a hyperbolic equation.}$$

$aaa - 6aa = 400$ $a = 10$

therefore $eee - 12e = +400$ (A)
$$+ 16$$

that is $eee - 12e = +416$, a hyperbolic equation.

$e = +8$

$a = e + r$

$a = 8 + 2 = 10$

---

2. $+3raa - aaa = -2xxx$

add $rrr - 3rra$ [to each side]

therefore $rrr - 3rra + 3raa - aaa = -2xxx - 3rra$
$$+ rrr$$

Let $r - a = e$

so $r - e = a$

$-3rra = -3rrr + 3rre$

therefore $eee - 3rre = -2xxx$
$$- 2rrr$$

therefore $+2xxx = +3rre - eee$
$$+ 2rrr$$

the conjugate equation to that in the first method, if that were possible, but it is not possible.[13]

Proof

Suppose $e = r$

therefore $\quad + 2xxx + 2rrr = + 3rrr - rrr$

$\quad\quad\quad\quad + 2xxx + 2rrr = + 2rrr$

therefore $\quad 2xxx = 0 \quad$ absurd

Suppose $\quad e > r$

then we will have $\quad e = r + f$

therefore $\quad + 2xxx + 2rrr = + 3rrr + 3rrf - rrr - 3rrf - 3rff - fff$

therefore $\quad + 2xxx = - 3rff - fff$

therefore $\quad + 2xxx + 3rff + fff = 0 \quad\quad$ absurd

Suppose $\quad e < r$

then we will have $\quad e + f = r$

and $\quad e = r - f$

therefore $\quad f < r$

therefore $\quad + 2xxx + 2rrr = + 3rrr - 3rrf - rrr + 3rrf - 3rff + fff$

$\quad\quad\quad\quad + 2xxx = + fff - 3rff = \dfrac{f - 3r}{ff}$

therefore $\quad f > 3r$

therefore $\quad f > r \quad\quad\quad$ absurd, because $f < r$

Therefore the conjugate equation is not possible.[14]

*e.*16) *On solving equations by reduction*

The equation to be reduced: $ccd = + baa + aaa$

or $2xxx = + 3raa + aaa$

I. $aaa + 3raa = 2xxx$

add $3rra + rrr$ [to each side]

therefore $aaa + 3raa + 3rra + rrr = \quad 2xxx + 3rra$
$$+ rrr$$

Let $a + r = e$

so $a = e - r$

$3rra = + 3rre - 3rrr$

therefore $eee - 3rre = + 2xxx$ (A)
$$- 2rrr$$

but if $2xxx < 2rrr$ it is possible to carry out *antithesis*

therefore $- 2xxx = + 3rre - eee$ (D)
$$+ 2rrr$$

---

2. $- 3raa - aaa = - 2xxx$

add $- rrr - 3rra$ [to each side]

therefore $- rrr - 3rra - 3raa - aaa = - 2xxx - 3rra$
$$- rrr$$

Let $- r - a = e$

so $- a = e + r$

and $- e - r = a$

$- 3rra = + 3rre + 3rrr$

therefore $eee - 3rre = - 2xxx$ (C) conjugate to (D)
$$+ 2rrr$$

but if $2xxx > 2rrr$ it is possible to carry out *antithesis*

therefore $+ 2xxx = + 3rre - eee$ (B) conjugate to (A)
$$- 2rrr$$

$aaa + 21aa = 972$ $\qquad\qquad$ $a = +6, -9, -18$

therefore $\quad eee - 147e = +972 = +147e - eee$ $\qquad$ (A), (B)
$$-686$$

that is $eee - 147e = +286 = +147e - eee$

$e = +13$	$e = +2$
$e = -2$	$e = +11$
$e = -11$	$e = -13$

$a = e - r$	$a = -e - r$
$a = +13 - 7 = 6$	$a = -2 - 7 = -9$
$a = -2 - 7 = -9$	$a = -11 - 7 = -18$
$a = -11 - 7 = -18$	$a = +13 - 7 = +6$

$aaa + 39aa = 3888$ $\qquad\qquad$ $a = +9, -12, -36$

therefore $\quad eee - 507e = \quad 3888 = +507e - eee$ $\qquad$ (C), (D)
$$+4394$$

that is $\quad eee - 507e = +506 = +507e - eee$

$e = +23$	$e = +1$
$e = -1$	$e = +22$
$e = -22$	$e = -23$

$a = -e - r$	$a = e - r$
$a = -23 - 13 = -36$	$a = +1 - 13 = -12$
$a = +1 - 13 = -12$	$a = +22 - 13 = +9$
$a = +22 - 13 = +9$	$a = -23 - 13 = -36$

Note:

It has already been noted that equations having conjugates are always ellip-tic, and they have negative as well as affirmative roots. The negative roots in one equation are the affirmative roots of the conjugate and conversely.[15] Which may be observed in section (*d*) on the generation of equations. Hyperbolic equations may not have conjugates and therefore have just one root.[16]

$$aaa + 21aa = 1856 \qquad\qquad a = 8$$
$$\text{therefore} \quad eee - 147e = + 1856 \qquad\qquad \text{(A)}$$
$$- 686$$
$$\text{that is} \quad eee - 147e = + 1170$$

a hyperbolic equation which does not have a conjugate;

nor, therefore, does the proposed equation.

$$e = 15$$
$$a = e - r$$
$$\text{therefore} \quad a = 15 - 7 = 8$$

e.17)   *On solving equations by reduction*

The equation to be reduced: $ccd = + baa - aaa$

or $2xxx = + 3raa - aaa$

---

1. $3raa - aaa = 2xxx$

add $rrr - 3rra$ [to each side]

therefore $rrr - 3rra + 3raa - aaa = + 2xxx - 3rra$

                                 $+ rrr$

Let $r - a = e$

so $r - e = a$

$- 3rra = - 3rrr + 3rre$

therefore $eee - 3rre = + 2xxx$ (A)

                       $- 2rrr$

but if $2xxx < 2rrr$ it is possible to carry out *antithesis*

therefore $- 2xxx = + 3rre - eee$ (D)

              $+ 2rrr$

---

2. $aaa - 3raa = - 2xxx$

add $+ 3rra - rrr$ [to each side]

therefore $aaa - 3raa + 3rra - rrr = - 2xxx + 3rra$

                                $- rrr$

Let $a - r = e$

so $a = e - r$

$3rra = + 3rre + 3rrr$

therefore $eee - 3rre = - 2xxx$ (C) conjugate to (D)

                     $+ 2rrr$

but if $2xxx > 2rrr$ it is possible to carry out *antithesis*

therefore $+ 2xxx = + 3rre - eee$ (B) conjugate to (A)

              $- 2rrr$

❧

$$972 = + 21aa - aaa \qquad\qquad a = +9, +18, -6$$

$$eee - 147e = +972 = +147e - eee \qquad\qquad \text{(A), (B)}$$
$$- 686$$

$$eee - 147e = +286 = +147e - eee$$

$e = +13$	$e = +2$
$e = -2$	$e = +11$
$e = -11$	$e = -13$

$a = r - e$	$a = e + r$
$a = 7 - 13 = -6$	$a = +2 + 7 = +9$
$a = 7 + 2 = +9$	$a = +11 + 7 = +18$
$a = 7 + 11 = +18$	$a = -13 + 7 = -6$

$$3888 = + 39aa - aaa \qquad\qquad a = +12, +36, -9$$

$$eee - 507e = -3888 = +507e - eee \qquad\qquad \text{(C), (D)}$$
$$+ 4394$$

$$eee - 507e = +506 = +507e - eee$$

$e = +23$	$e = +1$
$e = -1$	$e = +22$
$e = -22$	$e = -23$

$a = e + r$	$a = r - e$
$a = -23 + 13 = +36$	$a = +13 - 1 = +12$
$a = -1 + 13 = +12$	$a = +13 - 22 = -9$
$a = -22 + 13 = -9$	$a = +13 + 23 = +36$

In this case all the equations are elliptic, and none hyperbolic.

e.18)   *On solving equations by reduction*

The equation to be reduced: $ffg = + cda + baa + aaa$
or $2xxx = + ppa + 3raa + aaa$

---

1. $aaa + 3raa + ppa = 2xxx$

    $aaa + 3raa = + 2xxx - ppa$

add $3rra + rrr$ [to each side]

therefore $aaa + 3raa + 3rra + rrr = + 2xxx - ppa + 3rra$

                                            $+ rrr$

Let $a + r = e$

so $a = e - r$

$- ppa = - ppe + ppr$
$+ 3rra = + 3rre - 3rrr$

therefore $eee - 3rre = + 2xxx$ (A)

              $+ ppe$ $- 2rrr$

                  $+ ppr$

or $- 2xxx = + 3rre - eee$ (D)

    $+ 2rrr$ $- ppe$

    $- ppr$

---

2. $- aaa - 3raa - ppa = - 2xxx$

    $- aaa - 3raa = - 2xxx + ppa$

add $- 3rra - rrr$ [to each side]

therefore $- aaa - 3raa - 3rra - rrr = - 2xxx + ppa - 3rra$

                                              $- rrr$

Let $- a - r = e$

so $- e - r = a$

$+ ppa = - ppe - ppr$
$- 3rra = + 3rre + 3rrr$

therefore $eee - 3rre = - 2xxx$ (C) conjugate to (D)

              $+ ppe$ $+ 2rrr$

                  $- ppr$

or $+ 2xxx = + 3rre - eee$ (B) conjugate to (A)

    $- 2rrr$ $- ppe$

    $+ ppr$

$$aaa + 9aa + 8a = 60 \qquad\qquad a = +2, -5, -6$$

$$eee - 27e = +60 = +27e - eee \qquad\qquad \text{(A), (B)}$$
$$\quad +8e \quad -54 \quad -8e$$
$$\qquad\quad +24$$

$$eee - 19e = +30 = +19e - eee$$

$e = +5$	$e = +2$
$e = -2$	$e = +3$
$e = -3$	$e = -5$

$a = e - r$	$a = -e - r$
$a = +5 - 3 = +2$	$a = -2 - 3 = -5$
$a = -2 - 3 = -5$	$a = -3 - 3 = -6$
$a = -3 - 3 = -6$	$a = +5 - 3 = +2$

$$aaa + 27aa + 42a = 200 \qquad\qquad a = +2, -4, -25$$

$$eee - 243e = -200 \; = +243e - eee \qquad\qquad \text{(C), (D)}$$
$$\quad +42e \quad +1458 \quad -42e$$
$$\qquad\qquad -378$$

$$eee - 202e = +880 = +202e - eee$$

$e = +16$	$e = +5$
$e = -5$	$e = +11$
$e = -11$	$e = -16$

$a = -e - r$	$a = e - r$
$a = -16 - 9 = -25$	$a = +5 - 9 = -4$
$a = +5 - 9 = -4$	$a = +11 - 9 = +2$
$a = +11 - 9 = +2$	$a = -16 - 9 = -25$

$aaa + 9aa + 8a = 588$ $\qquad\qquad$ $a = 6$

$eee - 27e = +588$ $\qquad\qquad$ (A)
$ + 8e \quad -54$
$ +24$

$eee - 19e = +558$, a hyperbolic equation.

$e = 9$
$a = e - r$
$a = 9 - 3 = 6$

$aaa + 27aa + 255a = 626$ $\qquad\qquad$ $a = 2$

$eee - 243e = +626$ $\qquad\qquad$ (A)
$ +255e \quad -1458$
$ +2295$

$eee + 12e = +1463$

$e = 11$
$a = e - r$
$a = 11 - 9 = 2$

---

e.19)   *On solving equations by reduction*

The equation to be reduced: $ffg = -cda + baa - aaa$

or $2xxx = -ppa + 3raa - aaa$

---

1. $-ppa + 3raa - aaa = 2xxx$

$\quad 3raa - aaa = +2xxx + ppa$

add $rrr - 3rra$ [to each side]

therefore $rrr - 3rra + 3raa - aaa = +2xxx + ppa - 3rra$
$$\qquad\qquad\qquad\qquad\qquad\qquad\qquad + rrr$$

Let $r - a = e$

so $r - e = a$

$+ppa = +ppr - ppe$

$-3rra = -3rrr + 3rre$

therefore $eee - 3rre = +2xxx$ (A)
$$\qquad\qquad\quad + ppe \quad - 2rrr$$
$$\qquad\qquad\qquad\qquad\; + ppr$$

or $-2xxx = +3rre - eee$ (D)
$$\qquad\qquad + 2rrr \qquad\; - ppe$$
$$\qquad\qquad - ppr$$

---

2. $aaa - 3raa + ppa = -2xxx$

$\quad aaa - 3raa = -2xxx - ppa$

add $+3rra - rrr$ [to each side]

therefore $aaa - 3raa + 3rra - rrr = -2xxx - ppa + 3rra$
$$\qquad\qquad\qquad\qquad\qquad\qquad\qquad - rrr$$

Let $a - r = e$

so $a = e + r$

$-ppa = -ppe - ppr$

$+3rra = +3rre + 3rrr$

therefore $eee - 3rre = -2xxx$ (C) conjugate to (D)
$$\qquad\qquad\quad + ppe \quad + 2rrr$$
$$\qquad\qquad\qquad\qquad - ppr$$

or $+2xxx = +3rre - eee$ (B) conjugate to (A)
$$\; - 2rrr \qquad\; - ppe$$
$$\; + ppr$$

$$60 = -8a + 9aa - aaa \qquad\qquad a = +6, +5, -2$$

$$eee - 27e = +60 = +27e - eee \qquad\qquad \text{(A), (B)}$$
$$+8e \quad -54 \quad -8e$$
$$+24$$

$$eee - 19e = +30 = +19e - eee$$

$$e = +5 \qquad\qquad\qquad e = +2$$
$$e = -2 \qquad\qquad\qquad e = +3$$
$$e = -3 \qquad\qquad\qquad e = -5$$

$$a = r - e \qquad\qquad\qquad a = e + r$$
$$a = +3 - 5 = -2 \qquad\qquad a = +2 + 3 = +5$$
$$a = +3 + 2 = +5 \qquad\qquad a = +3 + 3 = +6$$
$$a = +3 + 3 = +6 \qquad\qquad a = -5 + 3 = -2$$

$$200 = -42a + 27aa - aaa \qquad\qquad a = +25, +4, -2$$

$$eee - 243e = -200 = +243e - eee \qquad\qquad \text{(C), (D)}$$
$$+42e \quad +1458 \quad -42e$$
$$-378$$

$$eee - 201e = +880 = +201e - eee$$

$$e = +16 \qquad\qquad\qquad e = +5$$
$$e = -5 \qquad\qquad\qquad e = +11$$
$$e = -11 \qquad\qquad\qquad e = -16$$

$$a = e + r \qquad\qquad\qquad a = r - e$$
$$a = +16 + 9 = +25 \qquad\qquad a = +9 - 5 = +4$$
$$a = -5 + 9 = +4 \qquad\qquad a = +9 - 11 = -2$$
$$a = -11 + 9 = -2 \qquad\qquad a = +9 + 16 = +25$$

---

*e.*20)   *On solving equations by reduction*

The equation to be reduced: $ffg = -cda + baa + aaa$

or $2xxx = -ppa + 3raa + aaa$

---

1. $aaa + 3raa - ppa = 2xxx$

      $aaa + 3raa = +2xxx + ppa$

add $3rra + rrr$ [to each side]

therefore $aaa + 3raa + 3rra + rrr = +2xxx + ppa + 3rra$

                                           $+ rrr$

Let $a + r = e$

so $a = e - r$

$+ ppa = +ppe - ppr$

$+ 3rra = +3rre - 3rrr$

therefore $eee - 3rre = +2xxx$ (A)

                  $- ppe$ $- 2rrr$

                      $- ppr$

or $-2xxx = +3rre - eee$ (D)

     $+2rrr$ $+ ppe$

     $+ ppr$

---

2. $+ ppa - 3raa - aaa = -2xxx$

      $- 3raa - aaa = -2xxx - ppa$

add $-rrr - 3rra$ [to each side]

therefore $-aaa - 3raa - 3rra - rrr = -2xxx - ppa - 3rra$

                                         $- rrr$

Let $-r - a = e$

so $-e - r = a$

$- ppa = +ppe + ppr$

$- 3rra = +3rre + 3rrr$

therefore $eee - 3rre = -2xxx$ (C) conjugate to (D)

                  $- ppe$ $+ 2rrr$

                      $+ ppr$

or $+2xxx = +3rre - eee$ (B) conjugate to (A)

     $- 2rrr$ $+ ppe$

     $- ppr$

$aaa + 6aa - 1a = 30$ $\qquad\qquad$ $a = +2, -3, -5$

$eee - 12e = +30 = +12e - eee$ $\qquad\qquad$ (A), (B)

$\qquad -1e \quad -16 \quad +1e$

$\qquad\qquad -12$

$eee - 13e = +12 = +13e - eee$

$e = +4$ $\qquad\qquad\qquad$ $e = +1$

$e = -1$ $\qquad\qquad\qquad$ $e = +3$

$e = -3$ $\qquad\qquad\qquad$ $e = -4$

$a = e - r$ $\qquad\qquad\qquad$ $a = -e - r$

$a = +4 - 2 = +2$ $\qquad\qquad$ $a = -1 - 2 = -3$

$a = -1 - 2 = -3$ $\qquad\qquad$ $a = -3 - 2 = -5$

$a = -3 - 2 = -5$ $\qquad\qquad$ $a = +4 - 2 = +2$

$aaa + 9aa - 12a = 20$ $\qquad\qquad$ $a = +2, -1, -10$

$eee - 27e = -20 = +27e - eee$ $\qquad\qquad$ (C), (D)

$\qquad -12e \quad +54 \quad +12e$

$\qquad\qquad +36$

$eee - 39e = +70 = +39e - eee$

$e = +7$ $\qquad\qquad\qquad$ $e = +2$

$e = -2$ $\qquad\qquad\qquad$ $e = +5$

$e = -5$ $\qquad\qquad\qquad$ $e = -7$

$a = -e - r$ $\qquad\qquad\qquad$ $a = e - r$

$a = -7 - 3 = -10$ $\qquad\qquad$ $a = +2 - 3 = -1$

$a = +2 - 3 = -1$ $\qquad\qquad$ $a = +5 - 3 = +2$

$a = +5 - 3 = +2$ $\qquad\qquad$ $a = -7 - 3 = -10$

$aaa + 6aa - 1a = 888$ $\qquad\qquad$ $a = 8$

$eee - 12e = +888$

$\qquad -1e \quad -16$

$\qquad\qquad -2$

$eee - 13e = +870$, a hyperbolic equation.

$e = 10$

$a = e - r$

$a = 10 - 2 = 8$

e.21)   *On solving equations by reduction*

The equation to be reduced: $ffg = + cda + baa - aaa$

or $2xxx = + ppa + 3raa - aaa$

I. $+ ppa + 3raa - aaa = 2xxx$

$\quad\quad 3raa - aaa = + 2xxx - ppa$

add $rrr - 3rra$ [to each side]

therefore $rrr - 3rra + 3raa - aaa = + 2xxx - ppa - 3rra$

$$+ rrr$$

Let $r - a = e$

so $r - e = a$

$- ppa = - ppr + ppe$

$- 3rra = - 3rrr + 3rre$

therefore $eee - 3rre = + 2xxx$ (A)

$\quad\quad\quad\quad\quad - ppe \quad - 2rrr$

$$- ppr$$

or $- 2xxx = + 3rre - eee$ (D)

$\quad + 2rrr \quad + ppe$

$\quad + ppr$

---

2. $aaa - 3raa - ppa = - 2xxx$

$\quad\quad aaa - 3raa = - 2xxx + ppa$

add $3rra - rrr$ [to each side]

therefore $aaa - 3raa + 3rra - rrr = - 2xxx + ppa + 3rra$

$$- rrr$$

Let $a - r = e$

so $a = e - r$

$+ ppa = + ppe + ppr$

$+ 3rra = + 3rre + 3rrr$

therefore $eee - 3rre = - 2xxx$ (C) conjugate to (A)

$\quad\quad\quad\quad\quad - ppe \quad + 2rrr$

$\quad\quad\quad\quad\quad\quad\quad + ppr$

or $+ 2xxx = + 3rre - eee$ (B) conjugate to (D)

$\quad - 2rrr \quad + ppe$

$\quad - ppr$

$30 = +1a + 6aa - aaa$ $\qquad\qquad$ $a = +3, +5, -2$

$eee - 12e = +30 = +12e - eee$ $\qquad\qquad$ (A), (B)

$\qquad -1e \quad -16 \quad +1e$

$\qquad\qquad -2$

$eee - 13e = +12 = +13e - eee$

$e = +4$	$e = +1$
$e = -1$	$e = +3$
$e = -3$	$e = -4$

$a = r - e$	$a = e + r$
$a = +2 - 4 = -2$	$a = +1 + 2 = +3$
$a = +2 + 1 = +3$	$a = +3 + 2 = +5$
$a = +2 + 3 = +5$	$a = -4 + 2 = -2$

$20 = +12a + 9aa - aaa$ $\qquad\qquad$ $a = +1, +10, -2$

$eee - 27e = -20 = +27e - eee$ $\qquad\qquad$ (C), (D)

$\qquad -12e \quad +54 \quad +12e$

$\qquad\qquad +36$

$eee - 39e = +70 = +39e - eee$

$e = +7$	$e = +2$
$e = -2$	$e = +5$
$e = -5$	$e = -7$

$a = e + r$	$a = r - e$
$a = +7 + 3 = +10$	$a = +3 - 2 = +1$
$a = -2 + 3 = +1$	$a = +3 - 5 = -2$
$a = -5 + 3 = -2$	$a = +3 + 7 = +10$

e.22)   *On solving equations by reduction*

The equation to be reduced: $ffg = -cda - baa + aaa$

or $2xxx = -ppa - 3raa + aaa$

---

1. $-ppa - 3raa + aaa = 2xxx$

       $aaa - 3raa = +2xxx + ppa$

add $3rra - rrr$ [to each side]

therefore $aaa - 3raa + 3rra - rrr = +2xxx + ppa + 3rra$

                                 $-rrr$

Let $a - r = e$

so $a = e + r$

$+ppa = +ppe + ppr$

$+3rra = +3rre + 3rrr$

therefore $eee - 3rre = +2xxx$ (A)

                  $-ppe$ $+2rrr$

                       $+ppr$

---

2. $+ppa + 3raa - aaa = -2xxx$

       $+3raa - aaa = -2xxx - ppa$

add $rrr - 3rra$ [to each side]

therefore $rrr - 3rra + 3raa - aaa = -2xxx - ppa - 3rra$

                                 $+rrr$

Let $r - a = e$

so $r - e = a$

$-ppa = -ppr + ppe$

$-3rra = -3rrr + 3rre$

therefore $eee - 3rre = -2xxx$

                  $-ppe$ $-2rrr$

                       $-ppr$

therefore $+2xxx = +3rre - eee$ (B) conjugate to (A)

            $+2rrr$ $+ppe$

            $+ppr$

$aaa - 3aa - 60a = 100$ $\qquad\qquad a = +10, -2, -5$

$eee - 3e = +100 = +3e - eee$ $\qquad\qquad$ (A), (B)
$\quad -60e \quad +2 \qquad +60e$
$\qquad\qquad +60$

$eee - 63e = +162 = +63e - eee$

$e = +9$	$e = +3$
$e = -3$	$e = +6$
$e = -6$	$e = -9$

$a = e + r$	$a = r - e$
$a = +9 + 1 = +10$	$a = +1 - 3 = -2$
$a = -3 + 1 = -2$	$a = +1 - 6 = -5$
$a = -6 + 1 = -5$	$a = +1 + 9 = +10$

$aaa - 3aa - 60a = 5600$ $\qquad\qquad a = 20$

$eee - 3e = +5600 \neq$ $\qquad\qquad$ (A)
$\quad -60e \quad +2$
$\qquad\qquad +60$

$eee - 63e = +5662$, a hyperbolic equation.

$e = 19$
$a = e + r$
$a = 19 + 1 = 20$

Add MS 6783 f. 190

*e.*23)   *On solving equations by reduction*

The equation to be reduced: $ffg = + cda - baa - aaa$
or $2xxx = + ppa - 3raa - aaa$

I. $+ ppa - 3raa - aaa = 2xxx$
     $- 3raa - aaa = + 2xxx - ppa$

add $- rrr - 3rra$ [to each side]
therefore $- rrr - 3rra - 3raa - aaa = + 2xxx - ppa - 3rra$
$$- rrr$$

Let $- r - a = e$
so $- r - e = a$

$- ppa = + ppr + ppe$
$- 3rra = + 3rrr + 3rre$

therefore $eee - 3rre = + 2xxx$ (A)
$$- ppe \quad + 2rrr$$
$$+ ppr$$

---

2. $- ppa + 3raa + aaa = - 2xxx$
     $+ 3raa + aaa = - 2xxx + ppa$

add $rrr + 3rra$ [to each side]
therefore $rrr + 3rra + 3raa + aaa = - 2xxx + ppa + 3rra$
$$+ rrr$$

Let $r + a = e$
so $a = e - r$

$+ ppa = + ppe - ppr$
$+ 3rra = + 3rre - 3rrr$

therefore $eee - 3rre = - 2xxx$
$$- ppe \quad - 2rrr$$
$$- ppr$$

therefore $+ 2xxx = + 3rre - eee$ (B) conjugate to (A)
$$+ 2rrr \quad + ppe$$
$$+ ppr$$

223

$$100 = +60a - 3aa - aaa \qquad\qquad a = +2, +5, -10$$

$$eee - 3e = +100 = +3e - eee \qquad\qquad \text{(A), (B)}$$
$$\quad -60e \quad +2 \qquad +60e$$
$$\qquad\qquad +60$$

$$eee - 63e = +162 = +63e - eee$$

$$e = +9 \qquad\qquad\qquad\qquad e = +3$$
$$e = -3 \qquad\qquad\qquad\qquad e = +6$$
$$e = -6 \qquad\qquad\qquad\qquad e = -9$$

$$a = -e - r \qquad\qquad\qquad\qquad a = e - r$$
$$a = -9 - 1 = -10 \qquad\qquad a = +3 - 1 = +2$$
$$a = +3 - 1 = +2 \qquad\qquad\; a = +6 - 1 = +5$$
$$a = +6 - 1 = +5 \qquad\qquad\; a = -9 - 1 = -10$$

*e.*24) *On solving equations by reduction*

The equation to be reduced: $ffg = + cda - baa + aaa$

or $2xxx = + ppa - 3raa + aaa$

I. $aaa - 3raa + ppa = 2xxx$

 $aaa - 3raa = + 2xxx - ppa$

add $+ 3rra - rrr$ [to each side]

therefore $aaa - 3raa + 3rra - rrr = + 2xxx - ppa + 3rra$
$$- rrr$$

Let $a - r = e$

so $a = e + r$

$- ppa = - ppe - ppr$

$+ 3rra = + 3rre + 3rrr$

therefore $eee - 3rre = + 2xxx$ (A)
$$+ ppe \quad + 2rrr$$
$$- ppr$$

or $- 2xxx = + 3rre - eee$ (D)
$$- 2rrr \quad - ppe$$
$$+ ppr$$

---

2. $- ppa + 3raa - aaa = - 2xxx$

 $3raa - aaa = - 2xxx + ppa$

add $rrr - 3rra$ [to each side]

therefore $rrr - 3rra + 3raa - aaa = - 2xxx + ppa - 3rra$
$$+ rrr$$

Let $r - a = e$

so $r - e = a$

$+ ppa = + ppr - ppe$

$- 3rra = - 3rrr + 3rre$

therefore $eee - 3rre = - 2xxx$ (C) conjugate to (D)
$$+ ppe \quad - 2rrr$$
$$+ ppr$$

or $+ 2xxx = + 3rre - eee$ (B) conjugate to (A)
$$+ 2rrr \quad - ppe$$
$$- ppr$$

$$aaa - 18aa + 87a = 110 \qquad\qquad a = +2, +5, +11$$

$$eee - 108e = +110 = +108e - eee \qquad\qquad \text{(A), (B)}$$
$$\quad +87e \qquad +432 \quad -87e$$
$$\qquad\qquad -522$$

$$eee - 21e = +20 = +21e - eee$$

$e = +5$	$e = +1$
$e = -1$	$e = +4$
$e = -4$	$e = -5$

$a = e + r$	$a = r - e$
$a = +5 + 6 = +11$	$a = +6 - 1 = +5$
$a = -1 + 6 = +5$	$a = +6 - 4 = +2$
$a = -4 + 6 = +2$	$a = +6 + 5 = +11$

$$aaa - 21aa + 126a = 176 \qquad\qquad a = +2, +8, +11$$

$$eee - 147e = -176 = +147e - eee \qquad\qquad \text{(C), (D)}$$
$$\quad +126e \quad -686 \quad -126e$$
$$\qquad\qquad +882$$

$$eee - 21e = +20 = +21e - eee \qquad\qquad \text{(as in the previous example)}$$

$e = +5$	$e = +1$
$e = -1$	$e = +4$
$e = -4$	$e = -5$

$a = r - e$	$a = e + r$
$a = 7 - 5 = +2$	$a = +1 + 7 = +8$
$a = 7 + 1 = +8$	$a = +4 + 7 = +11$
$a = 7 + 4 = +11$	$a = -5 + 7 = +2$

$aaa - 6aa + 11a = 12$ $\hspace{6cm}$ $a = 4$

$eee - 12e = +12$
$\hspace{1cm} + 11e \hspace{0.8cm} + 16$
$\hspace{2.5cm} - 22$
$eee - 1e = +6$, a hyperbolic equation.

$e = +2$
$a = e + r$
$a = +2 + 2 = 4$

$aaa - 6aa + 13a = 8$ $\hspace{6cm}$ $a = 1$

$eee - 12e = -8$
$\hspace{1cm} + 13e \hspace{0.8cm} - 16$
$\hspace{2.5cm} + 26$
$eee + 1e = +2$, a hyperbolic equation.

$e = +1$
$a = r - e$
$a = +2 - 1 = +1$

---

e.25)   *On solving equations by reduction*

The equation to be reduced: $ffg = -cda - baa - aaa$

or $2xxx = -ppa - 3raa - aaa$

---

1. $-ppa - 3raa - aaa = 2xxx$

       $-3raa - aaa = +2xxx + ppa$

add $-rrr - 3rra$ [to each side]

therefore $-rrr - 3rra - 3raa - aaa = +2xxx + ppa - 3rra$

$$-rrr$$

Let $-r - a = e$

so $-e - r = a$

$+ppa = -ppe - ppr$

$-3rra = +3rre + 3rrr$

therefore $eee - 3rre = +2xxx$ (A)

$$+ppe \qquad +2rrr$$

$$-ppr$$

or $-2xxx = +3rre - eee$ (D)

$$-2rrr \qquad -ppe$$

$$+ppr$$

---

2. $aaa + 3raa + ppa = -2xxx$

       $aaa + 3raa = -2xxx - ppa$

add $3rra + rrr$ [to each side]

therefore $aaa + 3raa + 3rra + rrr = -2xxx - ppa + 3rra$

$$+rrr$$

Let $a + r = e$

so $a = e - r$

$-ppa = -ppe + ppr$

$+3rra = +3rre - 3rrr$

therefore $eee - 3rre = -2xxx$ (C) conjugate to (D)

$$+ppe \qquad -2rrr$$

$$+ppr$$

or $+2xxx = +3rre - eee$ (B) conjugate to (A)

$$+2rrr \qquad -ppe$$

$$-ppr$$

---

e.26)   *On solving equations by reduction*

An appendix to clarify that place in Viète concerning the ambiguity of cubic equations affected more than once.[17] His book *De numerosa potestatum resolutione* page 31. The examples are explained. The precepts are amended. The lemmas are given and proved.

$aaa - 6aa + 11a = 6$ $a = +1, +2, +3$

$$eee - 12e = +6 \ = +12e - eee$$
$$ +11e \quad +16 \quad -11e$$
$$ -22$$
$$eee - 1e = +000 = +1e - eee$$

                                        (A), (B)

$eee = 1e$	$eee = 1e$
$ee \ = 1$	$ee \ = 1$
$e = +1$	$e = +0$
$e = -0$	$e = +1$
$e = -1$	$e = -1$
$a = e + r$	$a = r - e$
$a = +1 + 2 = +3$	$a = +2 - 0 = +2$
$a = -0 + 2 = +2$	$a = +2 - 1 = +1$
$a = -1 + 2 = +1$	$a = +2 + 1 = +3$

$aaa - 12aa + 29a = 18$ $a = +1, +2, +9$

$$eee - 48e = +18 \ = +48e - eee$$
$$ +29e \quad +128 \quad -29e$$
$$ -116$$
$$eee - 19e = +30 = +19e - eee$$

                                        (A), (B)

$e = +5$	$e = +2$
$e = -2$	$e = +3$
$e = -3$	$e = -5$
$a = e + r$	$a = r - e$
$a = +5 + 4 = +9$	$a = +4 - 2 = +2$
$a = -2 + 4 = +2$	$a = +4 - 3 = +1$
$a = -3 + 4 = +1$	$a = +4 + 5 = +9$

$$aaa - 18aa + 95a = 126 \qquad\qquad a = +2, +7, +9$$

$$eee - 108e = -126 = +108e - eee \qquad\qquad \text{(C), (D)}$$
$$+95e \quad\;\; -432 \quad -95e$$
$$+570$$

$$eee - 13e = +12 = +13e - eee$$

$e = +4$	$e = +1$
$e = -1$	$e = +3$
$e = -3$	$e = -4$

$a = r - e$	$a = e + r$
$a = +6 - 4 = +2$	$a = +1 + 6 = +7$
$a = +6 + 1 = +7$	$a = +3 + 6 = +9$
$a = +6 + 3 = +9$	$a = -4 + 6 = +2$

$$aaa - 9aa + 24a = 20 \qquad\qquad a = +2, +2, +5$$

$$eee - 27e = +20 = +27e - eee \qquad\qquad \text{(A), (B)}$$
$$+24e \quad +54 \quad -24e$$
$$-72$$

$$eee - 3e = +2 = +3e - eee \quad \text{a parabolic equation.}$$

$e = +2$	$e = +1$
$e = -1$	$e = +1$
$e = -1$	$e = -2$

$a = e + r$	$a = r - e$
$a = +2 + 3 = +5$	$a = +3 - 1 = +2$
$a = -1 + 3 = +2$	$a = +3 - 1 = +2$
$a = -1 + 3 = +2$	$a = +3 + 2 = +5$

$$aaa - 6aa + 12a = 8 \qquad\qquad\qquad a = +2, +2, +2$$

$$eee - 12e = +8 \;\; = +12e - eee \qquad\qquad\text{(A), (B)}$$
$$\quad\; +12e \quad -16 \quad -12e$$
$$\qquad\qquad -24$$

$$eee - 0e = +0 = +0e - eee$$

$$e = +0 \qquad\qquad\qquad e = +0$$
$$e = -0 \qquad\qquad\qquad e = +0$$
$$e = -0 \qquad\qquad\qquad e = -0$$

$$a = e + r \qquad\qquad\qquad a = r - e$$
$$a = +0 + 2 = +2 \qquad\qquad a = +2 - 0 = +2$$
$$a = -0 + 2 = +2 \qquad\qquad a = +2 - 0 = +2$$
$$a = -0 + 2 = +2 \qquad\qquad a = +2 + 0 = +2$$

That is, $\quad a = r.$

e.27)   *On solving equations by reduction*

Appendix

$$aaa - 6aa + 40a = 240 \qquad\qquad a = 6$$

$$eee - 12e = +240 \qquad\qquad\text{(A)}$$
$$+ 40e \quad + 16$$
$$- 80$$

$$eee + 28e = +176$$

$$e = +4$$

$$a = e + r$$
$$a = +4 + 2 = 6$$

But because $\begin{vmatrix} 6 \\ \underline{40} \end{vmatrix} = 240$ and $\begin{vmatrix} a \\ \underline{aa} \end{vmatrix} = aaa$

this is a reciprocal equation. Therefore the solution may be found by a short cut as is clear in (d.6).

$a =$ the coefficient of the square term, 6.

Viète's precepts amended.

$$aaa - 189aa + 4872a = 225790 \qquad\qquad a = 168$$

A cube affected by a negative square and a positive linear term is ambiguous when three times the square of a third of the coefficient of the square term is greater than the coefficient of the linear term (and the cube of a third of the coefficient of the square term is greater than the given constant).[18]

$aaa - 6aa + 11a = 6$	$a = 1,2,3$	(e.26)
$aaa - 6aa + 11a = 12$	$a = 4$	(e.24)

And when three times the square of a third of the coefficient of the square term is equal to the coefficient of the linear term (and the constant term plus twice the cube of that same third is equal to the product of a third of the coefficient of the square term and the coefficient of the linear term) there are three roots but a single equation.[19]

$$aaa - 6aa + 12a = 8 \qquad a = 2,2,2 \qquad (e.26)$$
$$aaa - 6aa + 12a = 9 \qquad a = 3$$

When three times the square (of a third) of the coefficient of the square term is less than the coefficient of the linear term, there is no ambiguity in the roots.[20]

$$aaa - 6aa + 13a = 8 \qquad a = 1 \qquad (e.24)$$

(Or when the cube of a third of the coefficient of the square term is less that the given constant.)[21]

$$aaa - 6aa + 11a = 12 \qquad a = 4 \qquad (e.24)$$
$$aaa - 6aa + 12a = 9 \qquad a = 3$$

But what need is there for verbose precepts, when with the formulae from our reduction, it is possible to show all the roots directly, not only for these cases, but for any other case you like. However, if a demonstration of these precepts is required, we adjoin the three following lemmas.

Lemma 1.
The canonical form for three roots: $(d.2)$, $(d.15)$, $(d.16)$

$$aaa - baa + bca \qquad a = b$$
$$\quad - caa + cda \qquad a = c$$
$$\quad - daa + bda = bcd \qquad a = d$$

e.28)   *On solving equations by reduction*

Appendix

Lemma 2, and proof

If a quantity is divided into three unequal parts, three times the square of a third of the sum of the parts is greater than [the sum of the three terms formed by taking each pair of unequal parts.

$$3, \frac{b+c+d}{3} \bigg| > bc + cd + bd$$
$$\frac{b+c+d}{3}$$

that is $\quad 3, \dfrac{b+c+d}{\underline{b+c+d}} \bigg| > bc + cd + bd$
$$\overline{9}$$

that is $\quad \dfrac{b+c+d}{\underline{b+c+d}} \bigg| > bc + cd + bd$
$$\overline{3}$$

therefore $\quad \dfrac{b+c+d}{b+c+d} \bigg| > 3bc + 3cd + 3bd$

that is $\quad bb + cc + dd + 2bc + 2cd + 2bd > 3bc + 3cd + 3bd$
that is $\quad bb + cc + dd > bc + cd + bd$
but $\quad bb + cc > 2bc \qquad$ (by lemma 1, e.8)
$$cc + dd > 2cd$$
$$bb + dd > 2bd$$

therefore $\quad 2bb + 2cc + 2dd > 2bc + 2cd + 2bd$
therefore $\quad bb + cc + dd > bc + cd + bd$

Therefore what must be proved is true.

e.29)   *On solving equations by reduction*

Appendix

Lemma 3, and proof

If a quantity is divided into three unequal parts, the cube of a third of the total is greater than the product of the three unequal parts.

$$\left.\begin{array}{c} \dfrac{b+c+d}{3} \\[2mm] \dfrac{b+c+d}{3} \\[2mm] \dfrac{b+c+d}{3} \end{array}\right| \quad > bcd$$

that is $\left.\begin{array}{c} b+c+d \\ b+c+d \\ b+c+d \\ \hline 27 \end{array}\right| \quad > bcd$

therefore $\left.\begin{array}{c} b+c+d \\ b+c+d \\ b+c+d \end{array}\right| \quad > 27bcd$

that is
$bbb + ccc + ddd + 3bcc + 3bdd + 3cbb + 3cdd + 3dbb + 3dcc + 6bcd > 27bcd$

therefore
$bbb + ccc + ddd + 3bcc + 3bdd + 3cbb + 3cdd + 3dbb + 3dcc > 21bcd$

but $\quad bbb + ccc > bbc + bcc \qquad$ (by lemma 2, *e*.8)
$\qquad ccc + ddd > ccd + cdd$
$\qquad bbb + ddd > bbd + bdd$

therefore $\quad 2bbb + 2ccc + 2ddd > bcc + bdd + cbb + cdd + dbb + dcc$

but $\quad bcc + bdd > 2bcd$

since $\quad \dfrac{cc + dd}{b} \biggr|\ \dfrac{> 2cd}{b}\biggr|$ $\quad$ (by lemma 1, *e*.8)

and $\quad cbb + cdd > 2bcd$

since $\quad \dfrac{bb + dd}{c}\biggr|\ \dfrac{> 2bd}{c}\biggr|$

and $\quad dbb + dcc > 2bcd$

since $\quad \dfrac{bb + cc}{d}\biggr|\ \dfrac{> 2cd}{d}\biggr|$

therefore $\quad bcc + bdd + cbb + cdd + dbb + dcc > 6bcd$

therefore $\quad 2bbb + 2ccc + 2ddd > 6bcd$

therefore $\quad bbb + ccc + ddd > 3bcd$

therefore $\quad 3bcc + 3bdd + 3cbb + 3cdd + 3dbb + 3dcc > 18bcd$

therefore $\quad bbb + ccc + ddd + 3bcc + 3bdd + 3cbb + 3cdd + 3dbb + 3dcc > 21bcd$

Therefore what must be proved is true.

§.

De resolutione aequationum per reductionem. (14.

Exempla antecedentis differentiâ in numeris.

xxxz + ffga − cd aa + Av aaa − aaaa.

$12 + 8a − 13aa + 8aaa − aaaa$. $a = 2.6$.

Hac aequatio non habet alias radices hypostaticas praeter 2, et 6. quoniam aequales sunt coefficienti longitudini.

Si essent plures, summa omnium esset maior. Quod est contra canones quatuor radicum.

Attamen duas alias habet noeticas (saluo canone) quas ex reductione sequenti, adgressu; confer exempli et ad ultimus contemplandum.

Reductio.   1° A.   2° B.

$+ zzzz − 24zz − 64z = −12 + 64z − 24zz + zzzz$.
$+ 13zz + 52z$ $+ 18$ $− 52z + 13zz$
$− 8z$ $− 52$ $+ 8z$
$+ 16$

$+ zzzz − 11zz − 20z = +00 = + 20z − 11zz + zzzz$

$z = −0$. $z = +0$.

Hyperb: $zzz − 11z = 20$. $20 + 11z − zzz$.

$z = 4$. $z = −4$.

radices istae ⎧ $z = 2 + \sqrt{−1}$
noeticae habet ⎨
per (2.13.)   ⎩ $z = 2 − \sqrt{−1}$

2°. $a = r − z$.

$a = 2 − 0 = 2$.
$a = 2 + 4 = 6$.
$a = 2, −2 − \sqrt{−1} = −\sqrt{−1}$. ⎫ noeticae
$a = 2 − 2 + \sqrt{−1} = +\sqrt{−1}$. ⎬ radices.

Æquationes quadrato-quadraticae quae habent duas radices aequales coefficienti longitudini generantur, (d. 13. 2°.)

On solving equations by reduction

───────────── ⌣ ─────────────

$f[8.1]$ *On solving equations by reduction*

The equations to be reduced: $xxxz = -2ppaa - aaaa$

$$xxxz = +2ppaa - aaaa$$

───────────────────────────────────

or $aaaa + 2ppaa = -xxxz = -2ppaa + aaaa$

add $+pppp$ [to each side]

therefore $aaaa + 2ppaa + pppp = +pppp - xxxz = +pppp - 2ppaa + aaaa$

───────────────────────────────────

1st $aa + pp = \sqrt{(pppp - xxxz)}$ 1st $pp - aa = \sqrt{(pppp - xxxz)}$

$\phantom{1^{st}}$ $aa = \sqrt{(pppp - xxxz)} - pp$ $\phantom{1^{st}}$ $pp - \sqrt{(pppp - xxxz)} = aa$

2nd $-aa - pp = \sqrt{(pppp - xxxz)}$ 2nd $aa - pp = \sqrt{(pppp - xxxz)}$

$\phantom{2^{nd}}$ $-\sqrt{(pppp - xxxz)} - pp = aa$ $\phantom{2^{nd}}$ $aa = pp + \sqrt{(pppp - xxxz)}$

───────────────────────────────────

Or   Or

1st $aa + pp = ee$ 1st $pp - aa = ee$

therefore $eeee = pppp - xxxz$ therefore $eeee = pppp - xxxz$

 $ee = \sqrt{(pppp - xxxz)}$ $ee = \sqrt{(pppp - xxxz)}$

 $aa = ee - pp$ $pp - ee = aa$

therefore $aa = -\sqrt{(pppp - xxxz)} - pp$ therefore $aa = pp - \sqrt{(pppp - xxxz)}$

2nd $-aa - pp = ee$ 2nd $aa - pp = ee$

therefore $eeee = pppp - xxxz$ therefore $eeee = pppp - xxxz$

 $ee = \sqrt{(pppp - xxxz)}$ $ee = \sqrt{(pppp - xxxz)}$

 $-ee - pp = aa$ $aa = pp + ee$

therefore $aa = -\sqrt{(pppp - xxxz)} - pp$ therefore $aa = pp + \sqrt{(pppp - sxxxz)}$

───────────────────────────────────

239

$f[\delta.2]$ *On solving equations by reduction*

The equations to be reduced: $xxxz = + 2ppaa + aaaa$

$$xxxz = - 2ppaa + aaaa$$

$$aaaa + 2ppaa = + xxxz = - 2ppaa + aaaa$$
$$\text{add } + pppp \text{ [to each side]}$$
therefore $aaaa + 2ppaa + pppp = + xxxz + pppp = + pppp - 2ppaa + aaaa$

1st $aa + pp = \sqrt{(xxxz + pppp)}$ 1st $pp - aa = \sqrt{(xxxz + pppp)}$
$\quad\quad aa = \sqrt{(xxxz + pppp)} - pp$ $\quad\quad pp - \sqrt{(xxxz + pppp)} = aa$

2nd $- aa - pp = \sqrt{(xxxz + pppp)}$ 2nd $aa - pp = \sqrt{(xxxz + pppp)}$
$\quad\quad - \sqrt{(xxxz + pppp)} - pp = aa$ $\quad\quad aa = pp + \sqrt{(xxxz + pppp)}$

Or   Or

1st $aa + pp = ee$ 1st $pp - aa = ee$
therefore $eeee = xxxz + pppp$ therefore $eeee = xxxz + pppp$
$\quad\quad\quad ee = \sqrt{(xxxz + pppp)}$ $\quad\quad\quad ee = \sqrt{(xxxz + pppp)}$
$\quad\quad\quad aa = ee - pp$ $\quad\quad\quad pp - ee = aa$
therefore $aa = \sqrt{(xxxz + pppp)} - pp$ therefore $aa = pp - \sqrt{(xxxz + pppp)}$

2nd $- aa - pp = ee$ 2nd $aa - pp = ee$
therefore $eeee = xxxz + pppp$ therefore $eeee = xxxz + pppp$
$\quad\quad\quad ee = \sqrt{(xxxz + pppp)}$ $\quad\quad\quad ee = \sqrt{(xxxz + pppp)}$
$\quad - ee - pp = aa$ $\quad\quad\quad aa = pp - ee$
therefore $aa = - \sqrt{(xxxz + pppp)} - pp$ therefore $aa = pp + \sqrt{(xxxz + pppp)}$

---

f[δ.3]   *On solving equations by reduction*

The equations to be reduced: $xxxz = -\,2ccda - aaaa$

$$xxxz = +\,2ccda - aaaa$$

---

or $aaaa + 2ccda = -\,xxxz = -\,2ccda + aaaa$

therefore $aaaa = \mp\,2ccda - xxxz$

$aaaa + ooaa + oooo = +\,ooaa \mp 2ccda - xxxz$
$\quad + ee[aa] + \dfrac{eeee}{4} \quad + ee[aa]^{1} \quad\quad + \dfrac{eeee}{4}$

$\left.\begin{array}{c} ee \\[4pt] \dfrac{eeee - xxxz}{4} \end{array}\right|$ $=$ $\left.\begin{array}{c} ccd \\[4pt] ccd \end{array}\right|$ $=\;\; ccdccd$

$\dfrac{eeeee - xxxzee}{4} = +ccdccd$

$eeeee - 4xxxzee = +\,2ccd2ccd$

therefore *ee* is given.[2]

$f[8.4]$ *On solving equations by reduction*

The equations to be reduced: $xxz = + 2ccda + aaaa$

$$xxz = - 2ccda + aaaa$$

---

or $aaaa + 2ccda = xxz = - 2ccda + aaaa$

therefore $aaaa = \mp 2ccda + xxz$

$$aaaa + ooaa + oooo = + ooaa \mp 2ccda + xxz$$
$$+ ee[aa] + \frac{eeee}{4} \quad + ee[aa] \qquad + \frac{eeee}{4}$$

$$\begin{array}{c|c|c} ee & ccd & \\ eeee + xxz & ccd & = ccdccd \\ \overline{4} & & \end{array}$$

$$\frac{eeeee + xxzee}{4} = + cccdccd$$

$$eeeee + 4xxzee = + 2ccd2ccd$$

therefore *ee* is given.

$f[8.5]$ *On solving equations by reduction*

The equations to be reduced: $xxxz = + 2ccda + 3ppaa + aaaa$

$$xxxz = - 2ccda + 3ppaa + aaaa$$

---

or $aaaa + 3ppaa + 2ccda = + xxxz = - 2ccda + 3ppaa + aaaa$

therefore $aaaa = - 3ppaa - 2ccda + xxxz$

$$aaaa + 00aa \quad + 0000 = - 3ppaa \mp 2ccda + xxxz$$

$$+ ee[aa] + \frac{eeee}{4} \qquad + ee[aa] \qquad + \frac{eeee}{4}$$

$$\left. \begin{array}{c} ee - 3pp \\ \dfrac{eeee + xxxz}{4} \end{array} \right| = \left. \begin{array}{c} ccd \\ ccd \end{array} \right| = \quad ccdccd$$

$$\frac{eeeee - 3ppeee}{4} \quad \frac{}{4} + xxxzee - 3ppxxxz = + cccdccd$$

$eeeee - 3ppeee + 4xxxzee - 3pp4xxxz = + 4ccdccd$

$eeeee - 3ppeee + 4xxxzee = + 2ccd2ccd + 3pp4xxxz$

$ee = + uu + pp$ $ee = - uu + pp$

$$uuuuuu - 3ppppuu = + 2ccd2ccd = + 3ppppuu - uuuuuu$$

$$+ 4xxxzuu \quad + 3pp4xxxz \quad - 4xxxzuu$$

$$+ 2pppppp$$

$$- 4xxxzpp$$

or

$$- uuuuuu + 3ppppuu = - 2ccd2ccd = - 3ppppuu + uuuuuu$$

$$- 4xxxzuu \quad - 3pp4xxxz \quad + 4xxxzuu$$

$$- 2pppppp$$

$$+ 4xxxzpp$$

$f[\delta.6]$ *On solving equations by reduction*

The equation to be reduced: $xxxz = -\ 2ccda - 3ppaa - aaaa$

$$xxxz = +\ 2ccda - 3ppaa - aaaa$$

---

or $aaaa + 3ppaa + 2ccda = -\ xxxz = -\ 2ccda + 3ppaa + aaaa$

therefore $aaaa = -\ 3ppaa - 2ccda - xxxz$

$$aaaa + 00aa\ + 0000 = -\ 3ppaa \mp 2ccda - xxxz$$

$$+\ ee[aa] + \frac{eeee}{4} \qquad +\ ee[aa] \qquad\qquad +\ \frac{eeee}{4}$$

$$\left.\begin{array}{c} ee - 3pp \\[4pt] \dfrac{eeee - xxxz}{4} \end{array}\right| = \left.\begin{array}{c} ccd \\[4pt] ccd \end{array}\right| = \quad ccdccd$$

$$\frac{eeeee}{4} - \frac{3ppeeee}{4} - xxxzee + 3ppxxxz = +\ cccdccd$$

$$eeeee - 3ppeeee - 4xxxzee + 3pp4xxxz = +\ 4ccdccd$$

$$eeeee - 3ppeeee - 4xxxzee = +\ 2ccd2ccd - 3pp4xxxz$$

$$ee = +\ uu + pp \qquad\qquad\qquad\qquad ee = -\ uu + pp$$

$$uuuuuu - 3ppppuu = +\ 2ccd2ccd\ = +\ 3ppppuu - uuuuuu$$

$$-\ 4xxxzuu \quad -\ 3pp4xxxz \quad +\ 4xxxzuu$$

$$+\ 2pppppp$$

$$+\ 4xxxzpp$$

or

$$-\ uuuuuu + 3ppppuu = -\ 2ccd2ccd\ = -\ 3ppppuu + uuuuuu$$

$$+\ 4xxxzuu \quad +\ 3pp4xxxz \quad -\ 4xxxzuu$$

$$-\ 2pppppp$$

$$-\ 4xxxzpp$$

$f[\delta.7]$ *On solving equations by reduction*

The equation to be reduced: $xxxz = -2ccda + 3ppaa - aaaa$

$$xxxz = +2ccda + 3ppaa - aaaa$$

---

or $aaaa - 3ppaa + 2ccda = -xxxz = -2ccda - 3ppaa + aaaa$

therefore $aaaa = +3ppaa - 2ccda - xxxz$

$$aaaa + ooaa + oooo = +3ppaa \mp 2ccda - xxxz$$
$$+ee[aa] + \frac{eeee}{4} \qquad +ee[aa] \qquad +\frac{eeee}{4}$$

$$\left. \begin{array}{l} ee + 3pp \\ \dfrac{eeee - xxxz}{4} \end{array} \right| = \left. \begin{array}{l} ccd \\ ccd \end{array} \right| = \quad ccdccd$$

$$eeeee + \frac{3ppeeee}{4} - xxxzee - \frac{3ppxxxz}{4} = +cccdccd$$

$$eeeee + 3ppeeee - 4xxxzee - 3pp4xxxz = +4ccdccd$$
$$eeeee + 3ppeeee - 4xxxzee = +2ccd2ccd - 3pp4xxxz$$

$$ee = +uu - pp \qquad\qquad\qquad ee = -uu - pp$$

$$\begin{array}{ll} uuuuuu - 3ppppuu = +2ccd2ccd & = +3ppppuu - uuuuuu \\ \quad - 4xxxzuu \quad + 3pp4xxxz & + 4xxxzuu \\ \qquad\qquad - 2pppppp \\ \qquad\qquad - 4xxxzpp \end{array}$$

or

$$\begin{array}{ll} -uuuuuu + 3ppppuu = -2ccd2ccd & = -3ppppuu + uuuuuu \\ \quad + 4xxxzuu \quad - 3pp4xxxz & - 4xxxzuu \\ \qquad\qquad + 2pppppp \\ \qquad\qquad + 4xxxzpp \end{array}$$

$\int[8.8]$ *On solving equations by reduction*

The equation to be reduced: $xxxz = + 2ccda - 3ppaa + aaaa$

$$xxxz = - 2ccda - 3ppaa + aaaa$$

---

or $aaaa - 3ppaa + 2ccda = xxxz = - 2ccda - 3ppaa + aaaa$

therefore $aaaa = + 3ppaa - 2ccda + xxxz$

$$aaaa + 00aa \; + 0000 = + 3ppaa \mp 2ccda - xxxz$$
$$+ ee[aa] + \underline{eeee} \;\;\;\;\; + ee[aa] \;\;\;\;\;\;\; + \underline{eeee}$$
$$\;\;\;\;\;\; 4 \; 4$$

$$\left. \begin{array}{c} \dfrac{eeee + xxxz}{4} \\ ee + 3pp \end{array} \right| = \left. \begin{array}{c} ccd \\ \overline{ccd} \end{array} \right| = ccdccd$$

$$\dfrac{eeeee + xxxzee + \underline{3ppeeee} + 3ppxxxz}{4 \;\;\;\;\;\;\;\;\;\;\;\;\;\;\; 4} = cccdccd$$

$$\dfrac{eeeee + \underline{3ppeeee} + xxxzee}{4 \;\;\;\;\;\; 4} = ccdccd - 3ppxxxz$$

$$eeeee + 3ppeeee + 4xxxzee = 4ccdccd - 12ppxxxz$$

$$ee = + uu - pp \; ee = - uu - pp$$

$$\begin{array}{llll} uuuuuu - & 3ppppuu = + 2ccd2ccd & = + 3ppppuu - & uuuuuu \\ & - 4xxxzuu & - 3pp4xxxz & - 4xxxzuu \\ & & - 2pppppp \\ & & + 4xxxzpp \end{array}$$

or

$$\begin{array}{llll} - uuuuuu + & 3ppppuu = - 2ccd2ccd & = - 3ppppuu + & uuuuuu \\ & - 4xxxzuu & + 3pp4xxxz & + 4xxxzuu \\ & & + 2pppppp \\ & & - 4xxxzpp \end{array}$$

Suppose $ee = ff$ so $\dfrac{eeee}{4} = \dfrac{ffff}{4}$

Let $3pp + ee = dd$ and $xxxz + \dfrac{eeee}{4} = cccc$

therefore $aaaa + ffaa + \dfrac{ffff}{4} = ddaa - 2cdda + cccc$

therefore 1st $aa + \dfrac{ff}{2} = da - cc$

2nd $aa + \dfrac{ff}{2} = cc - da$

$f[\alpha].1)$ *On solving equations by reduction*[3]

The equations to be reduced: $xxxz = + 4raaa + aaaa$

$$xxxz = - 4raaa + aaaa$$

$aaaa \pm 4raaa = xxxz$

add $+ 6rraa \pm 4rrra + rrrr$ [to each side]

therefore $aaaa \pm 4raaa + 6rraa \pm 4rrra + rrrr = \; + xxxz \pm 4rrra + 6rraa$

$$+ rrrr$$

1st Let $\pm a + r = e$

$$a = \pm e \mp r$$

$aa = ee - 2re + rr$

$\pm 4rrra = + 4rrre - 4rrrr$

$+ 6rraa = + 6rree - 12rrre + 6rrrr$

$eeee - 6rree + 8rrre = \; + xxxz$

$$+ 3rrrr$$

2nd Let $\mp a - r = e$

$$a = \mp e \mp r$$

$aa = ee + 2re + rr$

$\pm 4rrra = - 4rrre - 4rrrr$

$+ 6rraa = + 6rree + 12rrre + 6rrrr$

$+ xxxz = - 8rrre - 6rree + eeee$

$+ 3rrrr$

$r = 5$	$9261 = +20aaa + aaaa$	$a = +7, -21$
	$9261 = -20aaa + aaaa$	$a = -7, +21$

$$eeee - 150ee + 1000e = +9261 = -1000e - 150ee + eeee$$
$$+1875$$

that is:

$$eeee - 150ee + 1000e = +11136 = -1000e - 150ee + eeee$$

$e = +12$	$e = +16$
$e = -16$	$e = -12$

$a = e - r$	$a = -e - r$
$a = +12 - 5 = +7$	$a = -16 - 5 = -21$
$a = -16 - 5 = -21$	$a = +12 - 5 = +7$

$a = -e + r$	$a = +e + r$
$a = -12 + 5 = -7$	$a = +16 + 5 = +21$
$a = +16 + 5 = +21$	$a = -12 + 5 = -7$

Add MS 6783 f. 114

$f[\alpha].2)$ *On solving equations by reduction*

The equations to be reduced: $xxxz = -4raaa - aaaa$
$$xxxz = +4raaa - aaaa$$
contrary to the preceding equations.

$\pm a + r = e$	$\mp a - r = e$
$a = \pm e \mp r$	$a = \mp e \mp r$

$$eeee - 6rree + 8rrre = -xxxz = -8rrre - 6rree + eeee$$
$$+3rrrr$$

or

$$-eeee + 6rree - 8rrre = +xxxz = +8rrre + 6rree - eeee$$
$$-3rrrr$$

$r = 85$ $151732224 = -340aaa - aaaa$ $a = -84, -336$

$151732224 = +340aaa - aaaa$ $a = +84, +336$

$$eeee - 43350ee + 4913000e = -151732224 = -4913000e - 43350ee + eee$$
$$+ 156601875$$

that is:

$$eeee - 43350ee + 4913000e = +4869651 = -4913000e - 43350ee + eeee$$

$e = +1$	$e = -1$
$e = -251$	$e = +251$
$a = +e - r$	$a = -e - r$
$a = +1 - 85 = -84$	$a = +1 - 85 = -84$
$a = -251 - 85 = -336$	$a = -251 - 85 = -336$
$a = -e + r$	$a = +e + r$
$a = -1 + 85 = +84$	$a = -1 + 85 = +84$
$a = +251 + 85 = +336$	$a = +251 + 85 = +336$

$r = 10$ $59319 = -40aaa - aaaa$ $a = -13, -39$

$59319 = +40aaa - aaaa$ $a = +13, +39$

$$-eeee + 600ee - 8000e = +59319 = +8000e + 600ee - eeee$$
$$- 30000$$

that is:

$$-eeee + 600ee - 8000e = +29319 = +8000e + 600ee - eeee$$

$e = -3$	$e = +3$
$e = -29$	$e = +29$
$a = +e - r$	$a = -e - r$
$a = -3 - 10 = -13$	$a = -3 - 10 = -13$
$a = -29 - 10 = -39$	$a = -29 - 10 = -39$
$a = -e + r$	$a = +e + r$
$a = +3 + 10 = +13$	$a = +3 + 10 = +13$
$a = +29 + 10 = +39$	$a = +29 + 10 = +39$

$f[\alpha].3$)   *On solving equations by reduction*

The equation to be reduced: $xxxz = + ccda + 4raaa + aaaa$

$$xxxz = - ccda - 4raaa + aaaa$$

$aaaa \pm 4raaa \pm ccda = + xxxz$

$aaaa \pm 4raaa = + xxxz \mp ccda$

add $+6rraa \pm 4rrra + rrrr$ [to each side]

$aaaa \pm 4raaa + 6rraa \pm 4rrra + rrrr = + xxxz \mp ccda + 6rraa$
$$+ rrrr \pm 4rrra$$

1st Let $\pm a + r = e$

$\qquad a = \pm e \mp r$

$aa = ee - 2re + rr$

$\mp ccda \pm 4rrra = - ccde + ccdr + 4rrre - 4rrrr$

$+ 6rraa = + 6rree - 12rrre + 6rrrr$

$eeee - 6rree + 8rrre = + xxxz$

$\qquad\qquad + ccde \qquad + 3rrrr$

$\qquad\qquad\qquad\qquad + ccdr$

2nd Let $\mp a - r = e$

$\qquad a = \mp e \mp r$

$aa = ee + 2re + rr$

$\mp ccda \pm 4rrra = + ccde + ccdr - 4rrre - 4rrrr$

$+ 6rraa = + 6rree + 12rrre + 6rrrr$

$+ xxxz = - 8rrre - 6rree + eeee$

$+ 3rrrr \qquad - ccde$

$+ ccdr$

$r = \mathrm{I}$ $1024 = +128a + 4aaa + aaaa$ $a = +4, -8$

$1024 = -128a - 4aaa + aaaa$ $a = -4, +8$

$eeee - 6ee \quad +8e = +1024 = -8e - 6ee + eeee$

$\qquad +128e \quad +3 \qquad -128e$

$\qquad\qquad +128$

$eeee - 6ee + 136e = +1155 = -136e - 6ee + eeee$

$e = +5$ $e = +7$

$e = -7$ $e = -5$

$a = e - r$ $a = -e - r$

$a = +5 - \mathrm{I} = -4$ $a = -7 - \mathrm{I} = -8$

$a = -7 - \mathrm{I} = -8$ $a = +5 - \mathrm{I} = +4$

$a = -e + r$ $a = +e + r$

$a = -5 + \mathrm{I} = +4$ $a = +7 + \mathrm{I} = +8$

$a = +7 + \mathrm{I} = +8$ $a = -5 + \mathrm{I} = -4$

$f[\alpha].4)$ *On solving equations by reduction*

The equation to be reduced: $xxxz = + ppaa + 4raaa + aaaa$

$$xxxz = + ppaa - 4raaa + aaaa$$

$aaaa \pm 4raaa + ppaa = + xxxz$

$aaaa \pm 4raaa = + xxxz - ppaa$

add $+ 6rraa \pm 4rrra + rrrr$ [to each side]

$aaaa \pm 4raaa + 6rraa \pm 4rrra + rrrr = + xxxz \pm 4rrra - ppaa$

                                             $+ rrrr$ $+ 6rraa$

1st Let $\pm a + r = e$

          $a = \pm e \mp r$

$aa = ee - 2re + rr$

$\pm 4rrra = + 4rrre - 4rrrr$

$- ppaa + 6rraa = - ppee + 2ppre - pprr + 6ree - 12rrre + 6rrrr$

$eeee - 6ree + 8rrre \;\;\; = \;\; + xxxz$

     $+ ppee - 2ppre$ $+ 3rrrr$

                      $- pprr$

or

$-eeee + 6ree - 8rrre \;\;\; = \;\; - xxxz$

       $- ppee + 2ppre$ $- 3rrrr$

                        $+ pprr$

2nd Let $\mp a - r = e$

           $a = \mp e \mp r$

$aa = ee + 2re + rr$

$\pm 4rrra = - 4rrre - 4rrrr$

$- ppaa + 6rraa = - ppee - 2ppre - pprr + 6ree + 12rrre + 6rrrr$

$+ xxxz \;\; = \; - 8rrre - 6ree + eeee$

$+ 3rrrr \;\;\;\;\; + 2ppre + ppee$

$- pprr$

or

$- xxxz \;\; = \; + 8rrre + 6ree - eeee$

$- 3rrrr \;\;\;\;\; - 2ppre - ppee$

$+ pprr$

$f[\alpha].5)$ *On solving equations by reduction*

The foregoing equations: $xxxz = +ppaa + 4raaa + aaaa$

$$xxxz = +ppaa - 4raaa + aaaa$$

$r = 7$ $972 = +183aa + 28aaa + aaaa$ $a = +2, -3, -9, -18$

           $972 = +183aa - 28aaa + aaaa$ $a = -2, +3, +9, +18$

$eeee + 294ee - 2744e = -972 \quad = +2744e + 294ee - eeee$

       $-183ee + 2562e \quad\quad -7203 \quad\quad -2562e - 183ee$

                   $+8967$

$eeee + 111ee - 182e = +792 = +182e + 111ee - eeee$

$e = +4$	$e = +2$
$e = +9$	$e = +11$
$e = -2$	$e = -4$
$e = -11$	$e = -9$
$a = e - r$	$a = -e - r$
$a = +4 - 7 = -3$	$a = -2 - 7 = -9$
$a = +9 - 7 = +2$	$a = -11 - 7 = -18$
$a = -2 - 7 = -9$	$a = +4 - 7 = -3$
$a = -11 - 7 = -18$	$a = +9 - 7 = +2$
$a = -e + r$	$a = +e + r$
$a = -4 + 7 = +3$	$a = +2 + 7 = +9$
$a = -9 + 7 = -2$	$a = +11 + 7 = +18$
$a = +2 + 7 = +9$	$a = -4 + 7 = +3$
$a = +11 + 7 = +18$	$a = -9 + 7 = -2$

$r = 5$ $576 = + 100aa + 20aaa + aaaa$ $a = +2, -4, -6, -12$

$576 = +100aa - 20aaa + aaaa$ $a = -2, +4, +6, +12$

$eeee + 150ee - 1000e = -576 \quad = +1000e + 150ee - eeee$

$\quad -100ee + 1000e \quad\quad -1875 \quad -1000e - 100ee$

$\quad\quad\quad\quad +2500$

$eeee + 50ee\, (\pm\, 00e) = +49 = (\pm\, 00e) + 50ee - eeee$

$e = +1$	$e = +1$
$e = +7$	$e = +7$
$e = -1$	$e = -1$
$e = -7$	$e = -7$
$a = e - r$	$a = -e - r$
$a = +1 - 5 = -4$	$a = -1 - 5 = -6$
$a = +7 - 5 = +2$	$a = -7 - 5 = -12$
$a = -1 - 5 = -6$	$a = +1 - 5 = -4$
$a = -7 - 5 = -12$	$a = +7 - 5 = +2$
$a = -e + r$	$a = +e + r$
$a = -1 + 5 = +4$	$a = +1 + 5 = +6$
$a = -7 + 5 = -2$	$a = +7 + 5 = +12$
$a = +1 + 5 = +6$	$a = -1 + 5 = +4$
$a = +7 + 5 = +12$	$a = -7 + 5 = -2$

$f[\alpha].6)$ *On solving equations by reduction*

The equation to be reduced: $xxxz = + ccda + ppaa + 4raaa + aaaa$
$$xxxz = - ccda + ppaa - 4raaa + aaaa$$

$aaaa \pm 4raaa + ppaa \pm ccda = + xxxz$

$aaaa \pm 4raaa = + xxxz \mp ccda - ppaa$

add $+ 6rraa \pm 4rrra + rrrr$ [to each side]

$aaaa \pm 4raaa + 6rraa \pm 4rrra + rrrr = + xxxz \mp ccda - ppaa$
$$+ rrrr \pm 4rrra + 6rraa$$

1st Let $\pm a + r = e$

$\qquad a = \pm e \mp r$

$aa = ee - 2re + rr$

$\mp ccda \pm 4rrra = - ccde + ccdr + 4rrre - 4rrrr$

$- ppaa + 6rraa = - ppee + 2ppre - pprr + 6rree - 12rrre + 6rrrr$

$\begin{aligned} eeee - 6rree + 8rrre \quad &= \quad + xxxz \\ + ppee \;\; - 2ppre \quad & \quad + 3rrrr \\ + ccde \quad & \quad - pprr \\ & \quad + ccdr \end{aligned}$

or

$\begin{aligned} - eeee + 6rree - 8rrre \quad &= \quad - xxxz \\ - ppee \;\; + 2ppre \quad & \quad - 3rrrr \\ - ccde \quad & \quad + pprr \\ & \quad - ccdr \end{aligned}$

2nd Let $\mp a - r = e$

$$a = \mp e \mp r$$
$$aa = ee + 2re + rr$$

$\mp ccda \pm 4rrra = + ccde + ccdr - 4rrre - 4rrrr$
$- ppaa + 6rraa = - ppee - 2ppre - pprr + 6rree + 12rrre + 6rrrr$

$+ xxxz = - 8rrre - 6rree + eeee$
$+ 3rrrr \quad + 2ppre + ppee$
$- pprr \quad - ccde$
$+ ccdr$

or

$- xxxz = + 8rrre + 6rree - eeee$
$- 3rrrr \quad - 2ppre - ppee$
$+ pprr \quad + ccde$
$- ccdr$

$f[\alpha].7$)   *On solving equations by reduction*

The foregoing equations: $xxxz = + ccda + ppaa + 4raaa + aaaa$

$xxxz = - ccda + ppaa - 4raaa + aaaa$

$r = 3$ $48 = + 2a + 33aa + 12aaa + aaaa$ $a = + 1, - 2, - 3, - 8$

$48 = - 2a + 33aa - 12aaa + aaaa$ $a = - 1, + 2, + 3, + 8$

$eeee - 54ee + 216e = + 48 \ = \ - 216e - 54ee + eeee$

$+ 33ee - 198e \quad + 243 \quad + 198e + 33ee$

$+ 2e \qquad - 297 \quad - 2e$

$+ 6$

$eeee - 21ee + 20e = + 00 = - 20e - 21ee - eeee$

$e = + 0$ $e = - 0$

$- eee + 21e = 20$ $20e = 21ee + eee$

$e = + 1$ $e = + 5$

$e = + 4$ $e = - 1$

$e = - 5$ $e = - 4$

$a = e - r$ $a = - e - r$

$a = + 0 - 3 = - 3$ $a = + 0 - 3 = - 3$

$a = + 1 - 3 = - 2$ $a = - 5 - 3 = - 8$

$a = + 4 - 3 = + 1$ $a = + 1 - 3 = - 2$

$a = - 5 - 3 = - 8$ $a = + 4 - 3 = + 1$

$a = - e + r$ $a = + e + r$

$a = - 0 + 3 = + 3$ $a = - 0 + 3 = + 3$

$a = - 1 + 3 = + 2$ $a = + 5 + 3 = + 8$

$a = - 4 + 3 = - 1$ $a = - 1 + 3 = + 2$

$a = + 5 + 3 = + 8$ $a = - 4 + 3 = - 1$

$r = 4$ $\quad 72 = +6a + 49aa + 16aaa + aaaa \quad a = +1, -2, -3, -12$

$\qquad\qquad 72 = -6a + 49aa - 16aaa + aaaa \quad a = -1, +2, +3, +12$

$eeee - 96ee + 512e = +72 \quad = -512e - 96ee + eeee$

$\qquad + 49ee - 392e \qquad + 768 \qquad + 392e + 49ee$

$\qquad\qquad + 6e \qquad\qquad - 784 \qquad - 6e$

$\qquad\qquad\qquad\qquad + 24$

$eeee - 47ee + 126e = +80 = -126e - 47ee + eeee$

$e = +1$	$e = +8$
$e = +2$	$e = -1$
$e = +5$	$e = -2$
$e = -8$	$e = -5$

$a = e - r$	$a = -e - r$
$a = +1 - 4 = -3$	$a = -8 - 4 = -12$
$a = +2 - 4 = -2$	$a = +1 - 4 = -3$
$a = +5 - 4 = +1$	$a = +2 - 4 = -2$
$a = -8 - 4 = -12$	$a = +5 - 4 = +1$

$a = -e + r$	$a = +e + r$
$a = -1 + 4 = +3$	$a = +8 + 4 = +12$
$a = -2 + 4 = +2$	$a = -1 + 4 = +3$
$a = -5 + 4 = -1$	$a = -2 + 4 = +2$
$a = +8 + 4 = +12$	$a = -5 + 4 = -1$

$f[\gamma].8)$ *On solving equations by reduction*

The equation to be reduced: $hhhk = + ffga - cdaa + baaa - aaaa$
                              or: $xxxz = + ffga - cdaa + 4raaa - aaaa$

$xxxz = + ffga - cdaa + 4raaa - aaaa$

therefore $aaaa - 4raaa + cdaa - ffga = - xxxz$

therefore $aaaa - 4raaa = - xxxz + ffga - cdaa$

add $+ 6rraa - 4rrra + rrrr$ [to each side]

$aaaa - 4raaa + 6rraa - 4rrra + rrrr = - xxxz + ffga - cdaa$
$$+ rrrr - 4rrra + 6rraa$$

1st Let $a - r = e$

         $a = e + r$

$aa = ee + 2re + rr$

$+ ffga - 4rrra = + ffge + ffgr - 4rrre - 4rrrr$

$- cdaa + 6rraa = - cdee - 2cdre - cdrr + 6rree + 12rrre + 6rrrr$

$eeee - 6rree - 8rrre = - xxxz$ (A)
$\quad + cdee \ + 2cdre \ \ + 3rrrr$
$\quad\quad\quad - ffge \quad - cdrr$
$\quad\quad\quad\quad + ffgr$

or

$- eeee + 6rree + 8rrre = + xxxz$ (C)
$\quad - cdee \ - 2cdre \ - 3rrrr$
$\quad\quad\quad + ffge \quad + cdrr$
$\quad\quad\quad\quad - ffgr$

259

2nd Let $r - a = e$

$\qquad r - e = a$

$aa = rr - 2re + ee$

$+ ffga - 4rrra = + ffge - ffgr - 4rrrr + 4rrre$

$- cdaa + 6rraa = - cdrr + 2cdre - cdee + 6rrrr - 12rrre + 6rree$

$- xxxz = + 8rrre - 6rree + eeee \qquad\qquad\qquad\qquad \text{(B)}$

$+ 3rrrr \quad\ - 2cdre + cdee$

$- cdrr \quad\ + ffge$

$+ ffgr$

or

$+ xxxz = - 8rrre + 6rree - eeee \qquad\qquad\qquad\qquad \text{(D)}$

$- 3rrrr \quad\ + 2cdre - cdee$

$+ cdrr \quad\ - ffge$

$- ffgr$

$f[\gamma].9)$ *On solving equations by reduction*

Examples of the foregoing case in numbers

$$xxxz = + ffga - cdaa + 4raaa - aaaa$$

$672 = + 10004a - 371aa + 40aaa - aaaa$ $r = 11$ $a = 1, 3, 8, 28$

$eeee - 600ee - 8000e = -672$ $= +8000e - 600ee + eeee$ (A), (B)

$\quad\quad + 371ee + 7420e \quad +30000 \quad\quad -7420e + 371ee$

$\quad\quad\quad\quad -1004e \quad\quad -37100 \quad\quad +1004e$

$\quad\quad\quad\quad\quad\quad\quad\quad\quad +10040$

$eeee - 229ee - 1584e = +2268 = +1584e - 229ee + eeee$

$e = +18$	$e = +2$
$e = -2$	$e = +7$
$e = -7$	$e = +9$
$e = -9$	$e = -18$

1st $a = e + r$	2nd $a = r - e$
$a = +18 + 10 = +28$	$a = 10 - 2 = +8$
$a = -2 + 10 = +8$	$a = 10 - 7 = +3$
$a = -7 + 10 = +3$	$a = 10 - 9 = +1$
$a = -9 + 10 = +1$	$a = 10 + 18 = +28$

$3640 = +4362a - 769aa + 48aaa - aaaa$ $r = 12$ $a = 1, 13, 14, 20$

$eeee - 864ee - 13824e = -3640$ $= +13824e - 864ee + eeee$ (A), (B)

$\quad\quad +769ee + 18456e \quad +62208 \quad\quad -18456e + 769ee$

$\quad\quad\quad\quad -4362e \quad\quad -110736 \quad\quad +4362e$

$\quad\quad\quad\quad\quad\quad\quad\quad\quad +52344$

$eeee - 95ee + 270e = +176 = -270e - 95ee + eeee$

$e = +1$	$e = +11$
$e = +2$	$e = -1$
$e = +8$	$e = -2$
$e = -11$	$e = -8$

1st $a = e + r$	2nd $a = r - e$
$a = +1 + 12 = +13$	$a = 12 - 11 = +1$
$a = +2 + 12 = +14$	$a = 12 + 1 = +13$
$a = +8 + 12 = +20$	$a = 12 + 2 = +14$
$a = -11 + 12 = +1$	$a = 12 + 8 = +20$

Add MS 6783 f. 152

$f[\gamma].10)$ *On solving equations by reduction*

Examples of the foregoing case in numbers

$$xxxz = + ffga - cdaa + 4raaa - aaaa$$

$288 = + 444a - 179aa + 24aaaa - aaaa$ $r = 6$ $a = 1, 3, 8, 12$

$- eeee + 216ee + 1728e = + 288$	$= - 1728e + 216ee - eeee$	(C), (D)
$- 179ee - 2148e \quad - 3888$	$+ 2148e - 179ee$	
$+ 444e \quad\quad + 6444$	$- 444e$	
$- 2644$		

$- eeee + 37ee + 24e = + 180 = - 24e + 37ee - eeee$

$e = + 2$	$e = + 3$
$e = + 6$	$e = + 5$
$e = - 3$	$e = - 2$
$e = - 5$	$e = - 6$
$1^{st} \quad a = e + r$	$2^{nd} \quad a = r - e$
$a = + 2 + 6 = + 8$	$a = 6 - 3 = + 3$
$a = + 6 + 6 = + 12$	$a = 6 - 5 = + 1$
$a = - 3 + 6 = + 3$	$a = 6 + 2 = + 8$
$a = - 5 + 6 = + 1$	$a = 6 + 6 = + 12$

$360 = + 526a - 189aa + 24aaa - aaaa$ $r = 6$ $a = 1, 4, 9, 10$

$- eeee + 216ee + 1728e = + 360$	$= - 1728e + 216ee - eeee$	(C), (D)
$- 189ee - 2268e \quad - 3888$	$+ 2268e - 189ee$	
$+ 526e \quad\quad + 6804$	$- 526e$	
$- 3156$		

$- eeee + 27ee - 14e = + 120 = + 14e + 27ee - eeee$

$e = + 3$	$e = + 2$
$e = + 4$	$e = + 5$
$e = - 2$	$e = - 3$
$e = - 5$	$e = - 4$
$1^{st} \quad a = e + r$	$2^{nd} \quad a = r - e$
$a = + 3 + 6 = + 9$	$a = 6 - 2 = + 4$
$a = + 4 + 6 = + 10$	$a = 6 - 5 = + 1$
$a = - 2 + 6 = + 4$	$a = 6 + 3 = + 9$
$a = - 5 + 6 = + 1$	$a = 6 + 4 = + 10$

$f[\gamma].11)$ *On solving equations by reduction*

Examples of the foregoing case in numbers

$$xxxz = + ffga - cdaa + 4raaa - aaaa$$

$40 = + 78a - 49aa + 12aaa - aaaa$ $r = 3$ $a = 1, 2, 4, 5$

$- eeee + 54ee + 216e = + 40 \ = - 216e + 54ee - eeee$ (A), (B)

      $- 49ee - 294e \quad - 243 \quad + 294e - 49ee$

           $+ 78e \quad\quad + 441 \quad - 78e$

                 $- 234$

$- eeee + 5ee \, (\pm \, 0e) = + 4 = (\mp \, 0e) + 5ee - eeee$

$e = + 1$	$e = + 1$
$e = + 2$	$e = + 2$
$e = - 1$	$e = - 1$
$e = - 2$	$e = - 2$

1st	$a = e + r$	2nd	$a = r - e$
	$a = + 1 + 3 = + 4$		$a = 3 - 1 = + 2$
	$a = + 2 + 3 = + 5$		$a = 3 - 2 = + 1$
	$a = - 1 + 3 = + 2$		$a = 3 + 1 = + 4$
	$a = - 2 + 3 = + 1$		$a = 3 + 2 = + 5$

$$72 = + 134a - 77aa + 16aaa - aaaa \qquad r = 4 \qquad a = 1, 2, 4, 9$$

$$+ eeee - 96ee - 512e = -72 \quad = + 512e - 96ee + eeee \qquad \text{(A), (B)}$$
$$+ 77ee + 616e \quad + 768 \quad - 616e + 77ee$$
$$- 134e \quad - 1232 \quad + 134e$$
$$+ 536$$

$$+ eeee - 19ee - 30e = + 00 = + 30e - 19ee + eeee$$

$$eee - 19e = 30 \qquad\qquad e = + 0$$

$e = + 5$	$30 = + 19e - eee$
$e = - 0$	$e = + 2$
$e = - 2$	$e = + 3$
$e = - 3$	$e = - 5$

$1^{\text{st}} \quad a = e + r$ $\qquad\qquad$ $2^{\text{nd}} \quad a = r - e$

$a = + 5 + 4 = + 9$	$a = 4 - 0 = + 4$
$a = - 0 + 4 = + 4$	$a = 4 - 2 = + 2$
$a = - 2 + 4 = + 2$	$a = 4 - 3 = + 1$
$a = - 3 + 4 = + 1$	$a = 4 + 5 = + 9$

or $\quad - eeee + 19ee + 30e = + 00 = - 30e + 19ee - eeee$ $\qquad$ (C), (D)

$e = + 0$	$e = - 0$
$eee - 19e = 30$	$30 = + 19e - eee$
$e = + 5$	$e = + 2$
$e = - 2$	$e = + 3$
$e = - 3$	$e = - 5$
etc.	etc.

$f[\gamma].12)$ *On solving equations by reduction*

Examples of the foregoing case in numbers

$$xxxz = + ffga - cdaa + 4raaa - aaaa$$

$2560 = + 704a - 16aa + 4aaa - aaaa$ $r = 1$ $a = 4, 8$

$- eeee + 6ee \ + 8e \quad = + 2560 = \ - 8e + 6ee - eeee$ (C), (D)

$\qquad - 16ee - 32e \qquad - 3 \qquad + 32e - 16ee$

$\qquad\qquad + 704e \quad + 16 \qquad - 704e$

$\qquad\qquad\qquad\qquad - 704$

$- eeee - 10ee + 680e = + 1869 = - 680e - 10ee - eeee$

$e = + 3$ $e = - 3$

$e = + 7$ $e = - 7$

$1^{st} \quad a = e + r$ $2^{nd} \quad a = r - e$

$a = + 3 + 1 = + 4$ $a = 1 + 3 = + 4$

$a = + 7 + 1 = + 8$ $a = 1 + 7 = + 8$

$207 = + 312a - 120aa + 16aaa - aaaa$ $r = 4$ $a = 1, 3$

$+ eeee + \ 96ee + 512e = + 207 \ = - 512e + 96ee + eeee$ (C), (D)

$\qquad - 120ee - 960e \quad - 768 \qquad + 960e - 120ee$

$\qquad\qquad +312e \quad + 1920 \qquad - 312e$

$\qquad\qquad\qquad\qquad - 1248$

$eeee + 24ee - 136e = + 111 = + 136e - 24ee - eeee$

$e = - 1$ $e = + 1$

$e = - 3$ $e = + 3$

$1^{st} \quad a = e + r$ $2^{nd} \quad a = r - e$

$a = - 1 + 4 = + 3$ $a = 4 - 1 = + 3$

$a = - 3 + 4 = + 1$ $a = 4 - 3 = + 1$

$$33 = +56a - 30aa + 8aaa - aaaa \qquad\qquad r = 2 \qquad\qquad a = 1, 3$$

$$
\begin{aligned}
+ eeee - 24ee - 64e \;\; &= -33 \;\; = +64e - 24ee + eeee \qquad\qquad \text{(A), (B)}\\
+ 30ee + 120e \quad &+48 \qquad -120e + 30ee\\
-56e \qquad\; &-120 \quad +56e\\
&+112
\end{aligned}
$$

$$+ eeee + 6ee\,(-\,oe) = +7 = (+\,oe) + 6ee + eeee$$

$$
\begin{array}{ll}
e = +1 & \qquad\qquad e = +1\\
(e = -1) & \qquad\qquad (e = -1)\\[4pt]
1^{\text{st}} \;\; a = e + r & \qquad\qquad 2^{\text{nd}} \;\; a = r - e\\
a = +1 + 2 = +3 & \qquad\qquad a = 2 - 1 = +1\\
(a = -1 + 2 = +1) & \qquad\qquad (a = 2 + 1 = +3)
\end{array}
$$

Add MS 6783 f. 155

$f[\gamma].13)$ *On solving equations by reduction*

Examples of the foregoing case in numbers

$$xxxz = +ffga - cdaa + 4raaa - aaaa$$

$$44 = +67a - 30aa + 8aaa - aaaa \qquad\qquad r = 2 \qquad\qquad a = 1, 4$$

$$
\begin{aligned}
+ eeee - 24ee - 64e \;\; &= -44 \;\; = +64e - 24ee + eeee \qquad\qquad \text{(A), (B)}\\
+ 30ee + 120e \quad &+48 \qquad -120e + 30ee\\
-67e \qquad\; &-120 \quad +67e\\
&+134
\end{aligned}
$$

$$+ eeee + 6ee - 11e = +18 = +11e + 6ee + eeee$$

$$
\begin{array}{ll}
e = +2 & \qquad\qquad e = +1\\
e = -1 & \qquad\qquad e = -2\\[4pt]
1^{\text{st}} \;\; a = e + r & \qquad\qquad 2^{\text{nd}} \;\; a = r - e\\
a = +2 + 2 = +4 & \qquad\qquad a = 2 - 1 = +1\\
a = -1 + 2 = +1 & \qquad\qquad a = 2 + 2 = +4
\end{array}
$$

$$28 = + 31a - 6aa + 4aaa - aaaa \qquad\qquad r = 1 \qquad\qquad a = 1, 4$$

$$+ eeee - 6ee - 8e \;\; = - 28 = + 8e - 6ee + eeee \qquad\qquad \text{(A), (B)}$$
$$+ 6ee + 12e \quad + 3 \quad\; - 12e + 6ee$$
$$- 31e \quad\; - 6 \quad + 31e$$
$$+ 31$$

$$+ eeee \,(\mp 0ee) - 27e = + 00 = + 27e \,(\mp 0ee) + eeee$$
$$eeee = 27e$$
$$eee = 27$$

$e = 3$	$e = + 0$
$e = - 0$	$e = - 3$

1st $\quad a = e + r$	2nd $\quad a = r - e$
$a = + 3 + 1 = + 4$	$a = 1 + 0 = + 1$
$a = - 0 + 1 = + 1$	$a = 1 + 3 = + 4$

$$\text{or} \;-\; eeee \,(\pm 0ee) + 27e = + 00 = - 27e - eeee \qquad\qquad \text{(C), (D)}$$
$$eee = 27$$

$e = + 3$	$e = - 0$
$e = + 0$	$e = - 3$

1st $\quad a = e + r$	2nd $\quad a = r - e$
$a = + 3 + 1 = + 4$	$a = 1 + 0 = + 1$
$a = + 0 + 1 = + 1$	$a = 1 + 3 = + 4$

$\int[\gamma].14)^5$ *On solving equations by reduction*

Examples of the foregoing case in numbers

$$xxxz = + ffga - cdaa + 4raaa - aaaa$$

$12 = + 8a - 13aa + 8aaa - aaaa$ $a = 2, 6$

This equation has no other real roots,[6] besides 2 and 6, because their sum is equal to the coefficient of the cube term. If there were more, the sum of all of them would be greater. Which is against the canon for four roots.

However, (according to the canon) the equation has two other, imaginary, roots which I have shown in the following reduction, both as an example and as something that must be considered further.

The reduction:

$+ eeee - 24ee - 64e = -12 = +64e - 24ee + eeee$ (A), (B)

$\qquad + 13ee + 52e \qquad +48 \qquad -52e + 13ee$

$\qquad\qquad - 8e \qquad\qquad -52 \qquad +8e$

$\qquad\qquad\qquad\qquad +16$

$+ eeee - 11ee - 20e = +00 = +20e - 11ee + eeee$

$e = -0$ $e = +0$

$eee - 11e = 20$ $20 = +11e - eee$

(hyperbolic equation) $e = -4$

$e = 4$ these imaginary roots are to be had by (e.13):

$$e = 2 + \sqrt{-1}$$
$$e = 2 - \sqrt{-1}$$

$2^{nd} \quad a = r - e$

$a = 2 - 0 = 2$

$a = 2 + 4 = 6$

$\left. \begin{array}{l} a = 2 - 2 - \sqrt{-1} = -\sqrt{-1} \\ a = 2 - 2 + \sqrt{-1} = +\sqrt{-1} \end{array} \right\}$

imaginary roots

Biquadratic equations which have two roots whose sum is equal to the coefficient of the cube term may be generated as in (d. 13.2).[7]

---

f[γ].15)   *On solving equations by reduction*

Appendix
On finding certain equations in numbers, other than the foregoing

---

Lemma 1. If two or more equations of the same or different types have roots in common, the sum or difference of those equations will have the same roots. These kinds of equations I call cognates.

$$bcd = + bca - baa \qquad\qquad a = b$$
$$+ bda - caa \qquad\qquad a = c$$
$$+ cda - daa + aaa \qquad\qquad a = d$$
$$+ \qquad + \qquad +$$
$$bc = + ba - aa \qquad\qquad a = b$$
$$+ ca \qquad\qquad a = c$$

Lemma 2. If the terms of any equation are multiplied by any number, the equation so formed will have the same root or roots as the first. That is, the equations will be cognate.

$$bc = + ba \qquad\qquad a = b$$
$$+ ca - aa \qquad\qquad a = c$$

$$bcd = + bda \qquad\qquad a = b$$
$$+ cda - daa \qquad\qquad a = c$$

Thus $bca = + baa$ $\qquad\qquad a = b$
$$+ caa - aaa \qquad\qquad a = c$$

whence $000 = - bca + baa$ $\qquad\qquad a = b$
$$+ caa - aaa \qquad\qquad a = c$$

Examples of both lemmas in numbers.

$$6 = + 11a - 6aa + aaa \qquad\qquad a = 1, 2, 3$$
$$2 = + 3a - aa \qquad\qquad a = 1, 2$$
[add to give]
$$8 = + 14a - 7aa + aaa \qquad\qquad a = 1, 2, 4$$

$6 = + 11a - 6aa + aaa$	$a = 1, 2, 3$
$-2 = -3a + aa$	$a = 1, 2$
$\parallel$	
$4 = + 8a - 5aa + aaa$	$a = 1, 2$

$6 = + 11a - 6aa + aaa$	$a = 1, 2, 3$
$6 = + 6a$	$a = 1$
$\parallel$	
$12 = + 17a - 6aa + aaa$	$a = 1$

$2 = + 31 - aa \,\big|$	$a = 1, 2$
$ 2$	

$4 = + 6a - 2aa$	$a = 1, 2$

Add MS 6783 f. 159

$\int[\gamma].16)$ *On solving equations by reduction*

Appendix

Problem

Suppose an equation has the given form:

$$xxxz = + ffga - cdaa + 4raa - aaaa$$

Potentially it has four roots. It is required to find an equation of the same form in numbers that has only two roots. Because the fourth degree term is negative it cannot have one or three [positive] roots.

$39 = + 40a - aaaa$	$a = 1, 3$
$12 = + 13a - aaa$	$a = 1, 3$
$3 = + 4a - aa$	$a = 1, 3$

$-12 = -13a + aaa$ $\qquad\qquad\quad 8$	$a = 1, 3$
$-96 = -104a + 8aaa$	$a = 1, 3$
$+39 = +40a - aaaa$	$a = 1, 3$
$\parallel$	
$-57 = -64a + 8aaa - aaaa$	$a = 1, 3, X$
$3 = +4a - aa$ $\qquad\quad 30$	$a = 1, 3$
$90 = +120a - 30aa$	$a = 1, 3$
$-57 = -64a + 8aaa - aaaa$	$a = 1, 3, X$
$\parallel$	
$33 = +56a - 30aa + 8aaa - aaaa$	$a = 1, 3$
the required equation.	

The problem more briefly solved

$40 = +78a - 49aa + 12aaa - aaaa$	$a = 1, 2, 4, 5$
$60 = +90a - 30aa$	$a = 1, 2$
$\parallel$	
$100 = +168a - 79aa + 12aaa - aaaa$	$a = 1, 2$
the required equation, for *rrrr* < *xxxz* that is	$3 \,\big|\, = 81 < 100$ $3 \,\big|$ $3 \,\big|$ $3 \,\big|$ etc.

$f[\gamma].17)$ *On solving equations by reduction*

The equation to be reduced: $hhhk = + ffga + cdaa - baaa + aaaa$

or: $xxxz = + ffga + cdaa - 4raaa + aaaa$

$xxxz = + ffga + cdaa - 4raaa + aaaa$

also $aaaa - 4raaa + cdaa + ffga = + xxxz$

therefore $aaaa - 4raaa = + xxxz - ffga - cdaa$

add $+ 6rraa - 4rrra + rrrr$ [to each side]

$aaaa - 4raaa + 6rraa - 4rrra + rrrr = + xxxz - ffga - cdaa$

$+ rrrr - 4rrra + 6rraa$

1st Let $a - r = e$

$a = e + r$

$aa = ee + 2re + rr$

$- ffga - 4rrra = - ffge - ffgr - 4rrre - 4rrrr$

$- cdaa + 6rraa = - cdee - 2cdre - cdrr + 6rree + 12rrre + 6rrrr$

$eee - 6rree - 8rrre \ = + xxxz$ (A)

$+ cdee \ + 2cdre \quad + 3rrrr$

$+ ffge \qquad - cdrr$

$- ffgr$

or

$- eeee + 6rree + 8rrre \ = - xxxz$ (C)

$- cdee \ - 2cdre \quad - 3rrrr$

$- ffge \qquad + cdrr$

$+ ffgr$

272

2nd Let $r - a = e$

$\qquad r - e = a$

$aa = rr - 2re + ee$

$- ffga - 4rrra = - ffgr + ffge - 4rrrr + 4rrre$

$- cdaa + 6rraa = - cdrr + 2cdre - cdee + 6rrrr - 12rrre + 6rree$

$+ xxxz \;\; = + 8rrre - 6rree + eeee$ $\qquad\qquad$ (B)

$+ 3rrrr \qquad - 2cdre + cdee$

$- cdrr \qquad\;\; - ffge$

$- ffgr$

or

$- xxxz \;\; = - 8rrre + 6rree - eeee$ $\qquad\qquad$ (D)

$- 3rrrr \qquad + 2cdre - cdee$

$+ cdrr \qquad\;\; + ffge$

$+ ffgr$

Add MS 6783 f. 162

$f[\gamma].18$)   *On solving equations by reduction*

Examples of the foregoing case in numbers

$$xxxz = + ffga + cdaa - 4raaa + aaaa$$

$6 = +8a + \text{I}aa - 4aaa + aaaa$ $r = \text{I}$ $a = \text{I}, 3, \sqrt{2}, -\sqrt{2}$

$+ eeee - 6ee - 8e = +6 = +8e - 6ee + eeee$ (A), (B)

     $+ \text{I}ee + 2e$ $+3$ $-2e + \text{I}ee$

         $+8e$ $-\text{I}$ $-8e$

               -8

$+ eeee - 5ee + 2e = +00 = -2e - 5ee + eeee$

$e = +0$	$e = -0$
$2 = +5e - eee$	$2 = -5e + eee$
$e = +2$	$e = +\sqrt{2} + \text{I}$
$e = +\sqrt{2} - \text{I}$	$e = -2$
$e = -\sqrt{2} - \text{I}$	$e = -\sqrt{2} - \text{I}$

I^{st} $a = e + r$ 2nd $a = r - e$

$a = +0 + \text{I} = +\text{I}$ $a = \text{I} + 0 = +\text{I}$

$a = +2 + \text{I} = +3$ $a = \text{I} - \sqrt{2} - \text{I} = +\sqrt{2}$

$a = +\sqrt{2} - \text{I} + \text{I} = +\sqrt{2}$ $a = \text{I} + 2 = +3$

$a = -\sqrt{2} - \text{I} + \text{I} = -\sqrt{2}$ $a = \text{I} - \sqrt{2} - \text{I} = -\sqrt{2}$

$$30 = + 16a + 13aa - 8aaa + aaaa \qquad\qquad r = 2 \qquad\qquad a = 3, 5, \sqrt{2}, -\sqrt{2}$$

$$- eeee + 24ee + 64e = - 30 = - 64e + 24ee - eeee \qquad \text{(C), (D)}$$
$$- 13ee - 52e \quad - 48 \quad + 52e - 13ee$$
$$- 16e \quad + 52 \quad + 16e$$
$$+ 32$$

$$- eeee + 11ee - 4e = + 6 = + 4e + 11ee - eeee$$

$e = + 1$	$e = + 2 + \sqrt{2}$
$e = + 3$	$e = + 2 - \sqrt{2}$
$e = - \sqrt{2} - 2$	$e = - 1$
$e = - 2 + \sqrt{2}$	$e = - 3$

$1^{st} \quad a = e + r$	$2^{nd} \quad a = r - e$
$a = + 1 + 2 = + 3$	$a = + 2 - 2 - \sqrt{2} = - \sqrt{2}$
$a = + 3 + 2 = + 5$	$a = + 2 - 2 + \sqrt{2} = + \sqrt{2}$
$a = - \sqrt{2} - 2 + 2 = - \sqrt{2}$	$a = + 2 + 1 = + 3$
$a = - 2 + \sqrt{2} + 2 = + \sqrt{2}$	$a = + 2 + 3 = + 5$

In these examples the sum of the two [integer] roots is equal to the coefficient of the cube term, and thus also to the sum of all four roots. This type of equation may be generated as in (d.13.2).

$f[ζ.1])$ *On solving equations by reduction*

The equations to be reduced:
$$xxxz = -ccda - ppaa - 4raaa - aaaa$$
$$xxxz = +ccda - ppaa + 4raaa - aaaa$$

$$aaaa \pm 4raaa + ppaa \pm ccda = -xxxz$$
$$aaaa \pm 4raaa = -xxxz \mp ccda - ppaa$$
add $+6rraa \pm 4rrra + rrrr$ [to each side]

$$aaaa \pm 4raaa + 6rraa \pm 4rrra + rrrr = -xxxz \mp ccda - ppaa$$
$$+ rrrr \pm 4rrra + 6rraa$$

1st Let $\pm a + r = e$
$$a = \pm e \mp r$$
$$aa = +ee - 2re + rr$$

$$\mp ccda \pm 4rrra = -ccde + ccdr + 4rrre - 4rrrr$$
$$-ppaa + 6rraa = -ppee + 2ppre - pprr + 6rree - 12rrre + 6rrrr$$

$$eeee - 6rree + 8rrre \;=\; -xxxz$$
$$+ ppee \;\; - 2ppre \quad + 3rrrr$$
$$+ ccde \qquad\quad - pprr$$
$$+ ccdr$$

or

$$-eeee + 6rree - 8rrre \;=\; +xxxz$$
$$- ppee \;\; + 2ppre \quad - 3rrrr$$
$$- ccde \qquad\quad + pprr$$
$$- ccdr$$

2nd Let $\mp a - r = e$

$$a = \mp e \mp r$$
$$aa = + ee + 2re + rr$$

$$\mp ccda \pm 4rrra = + ccde + ccdr - 4rrre - 4rrrr$$
$$- ppaa + 6rraa = - ppee - 2ppre - pprr + 6rree + 12rrre + 6rrrr$$

$$
\begin{aligned}
- xxxz &= - 8rrre - 6rree + eeee \\
+ 3rrrr &\quad + 2ppre + ppee \\
- pprr &\quad - ccde \\
+ ccdr
\end{aligned}
$$

or

$$
\begin{aligned}
+ xxxz &= + 8rrre + 6rree - eeee \\
- 3rrrr &\quad - 2ppre - ppee \\
+ pprr &\quad + ccde \\
- ccdr
\end{aligned}
$$

Corollary

It thus appears that all these equations are the same as their foregoing pairs of conjugates [see $(f[\alpha].6)$], which are contrary to them, except that $xxxz$ has the opposite sign, and so it is in all contrary equations. In what follows, therefore, it can be put more concisely.

Add MS 6783 f. 200

$f[\varsigma.2])$ *On solving equations by reduction*

The equations to be reduced: $xxxz = + ccda - ppaa + 4raaa + aaaa$

$$xxxz = - ccda - ppaa - 4raaa + aaaa$$
$$xxxz = - ccda + ppaa - 4raaa - aaaa$$
$$xxxz = + ccda + ppaa + 4raaa - aaaa$$

$aaaa \pm 4raaa + ppaa \pm ccda = + xxxz$

$aaaa \pm 4raaa = + xxxz \mp ccda + ppaa$

add $+ 6rraa \pm 4rrra + rrrr$ [to each side]

$aaaa \pm 4raaa + 6rraa \pm 4rrra + rrrr = + xxxz \mp ccda + ppaa$
$$+ rrrr \pm 4rrra + 6rraa$$

1st Let $\pm a + r = e$

       $a = \pm e \mp r$

$aa = + ee - 2re + rr$

$\mp ccda \pm 4rrra = - ccde + ccdr + 4rrre - 4rrrr$

$+ ppaa + 6rraa = + ppee - 2ppre + pprr + 6rree - 12rrre + 6rrrr$

$eeee - 6rree + 8rrre \ = \ \pm xxxz$

     $- ppee \ + 2ppre \ \ \ \ + 3rrrr$

            $+ ccde \ \ \ \ \ \ \ + pprr$

                  $+ ccdr$

or

$- eeee + 6rree - 8rrre \ = \ \mp xxxz$

     $+ ppee \ - 2ppre \ \ \ \ - 3rrrr$

            $- ccde \ \ \ \ \ \ \ - pprr$

                  $- ccdr$

2nd Let $\mp a - r = e$

$$a = \mp e \mp r$$
$$aa = +ee + 2re + rr$$

$$\mp ccda \pm 4rrra = +ccde + ccdr - 4rrre - 4rrrr$$
$$+ ppaa + 6rraa = +ppee + 2ppre + pprr + 6rree + 12rrre + 6rrrr$$

$$\pm xxxz = -8rrre - 6rree + eeee$$
$$+ 3rrrr \quad - 2ppre - ppee$$
$$+ pprr \quad - ccde$$
$$+ ccdr$$

or

$$\mp xxxz = +8rrre + 6rree - eeee$$
$$- 3rrrr \quad + 2ppre + ppee$$
$$- pprr \quad + ccde$$
$$- ccdr$$

$f[\varsigma.3])$ *On solving equations by reduction*

The equations to be reduced:

$$xxxz = -ccda + ppaa + 4raaa + aaaa$$
$$xxxz = +ccda + ppaa - 4raaa + aaaa$$
$$xxxz = +ccda - ppaa - 4raaa - aaaa$$
$$xxxz = -ccda - ppaa + 4raaa - aaaa$$

$aaaa \pm 4raaa + ppaa \mp ccda = +xxxz$

$aaaa \pm 4raaa = +xxxz \pm ccda - ppaa$

add $+6rraa \pm 4rrra + rrrr$ [to each side]

$aaaa \pm 4raaa + 6rraa \pm 4rrra + rrrr = +xxxz \pm ccda - ppaa$
$$+ rrrr \pm 4rrra + 6rraa$$

1st Let $\pm a + r = e$

$\quad\quad a = \pm e \mp r$

$aa = +ee - 2re + rr$

$\pm ccda \pm 4rrra = +ccde - ccdr + 4rrre - 4rrrr$

$-ppaa + 6rraa = -ppee + 2ppre - pprr + 6rree - 12rrre + 6rrrr$

$eeee - 6rree + 8rrre \ = \ \pm xxxz$
$\quad + ppee \ \ - 2ppre \quad + 3rrrr$
$\quad\quad\quad\quad - ccde \quad\quad - pprr$
$\quad\quad\quad\quad\quad\quad\quad\quad - ccdr$

or

$-eeee + 6rree - 8rrre \ = \ \mp xxxz$
$\quad\quad - ppee \ \ + 2ppre \quad - 3rrrr$
$\quad\quad\quad\quad + ccde \quad\quad + pprr$
$\quad\quad\quad\quad\quad\quad\quad\quad + ccdr$

2nd Let $\mp a - r = e$

$\qquad\qquad a = \mp e \mp r$

$aa = + ee + 2re + rr$

$\pm ccda \pm 4rrra = - ccde - ccdr - 4rrre - 4rrrr$

$- ppaa + 6rraa = - ppee - 2ppre - pprr + 6rree + 12rrre + 6rrrr$

$+ xxxz = - 8rrre - 6rree + eeee$

$+ 3rrrr \quad + 2ppre + ppee$

$- pprr \quad + ccde$

$- ccdr$

or

$- xxxz = + 8rrre + 6rree - eeee$

$- 3rrrr \quad - 2ppre - ppee$

$+ pprr \quad - ccde$

$+ ccdr$

$f[ζ.4])$ *On solving equations by reduction*

The equations to be reduced: $xxxz = -ccda - ppaa + 4raaa + aaaa$
$$xxxz = +ccda - ppaa - 4raaa + aaaa$$
$$xxxz = +ccda + ppaa - 4raaa - aaaa$$
$$xxxz = -ccda + ppaa + 4raaa - aaaa$$

$aaaa \pm 4raaa - ppaa \mp ccda = +xxxz$

$aaaa \pm 4raaa = +xxxz \pm ccda + ppaa$

add $+6rraa \pm 4rrra + rrrr$ [to each side]

$aaaa \pm 4raaa + 6rraa \pm 4rrra + rrrr = +xxxz \pm ccda + ppaa$
$$+ rrrr \pm 4rrra + 6rraa$$

1st Let $\pm a + r = e$

$\qquad a = \pm e \mp r$

$aa = + ee - 2re + rr$

$\pm ccda \pm 4rrra = +ccde - ccdr + 4rrre - 4rrrr$

$+ ppaa + 6rraa = + ppee - 2ppre + pprr + 6rree - 12rrre + 6rrrr$

$eeee - 6rree + 8rrre\ = +xxxz$
$\qquad - ppee\ + 2ppre\quad + 3rrrr$
$\qquad\qquad - ccde\qquad + pprr$
$\qquad\qquad\qquad - ccdr$

or

$- eeee + 6rree - 8rrre\ = -xxxz$
$\qquad + ppee\ - 2ppre\qquad - 3rrrr$
$\qquad\qquad + ccde\qquad - pprr$
$\qquad\qquad\qquad + ccdr$

2nd Let $\mp a - r = e$

$$a = \mp e \mp r$$
$$aa = + ee + 2re + rr$$

$\pm ccda \pm 4rrra = - ccde - ccdr - 4rrre - 4rrrr$

$+ ppaa + 6rraa = + ppee + 2ppre + pprr + 6rree + 12rrre + 6rrrr$

$+ xxxz = - 8rrre - 6rree + eeee$

$+ 3rrrr \quad - 2ppre - ppee$

$+ pprr \quad + ccde$

$- ccdr$

or

$- xxxz = + 8rrre + 6rree - eeee$

$- 3rrrr \quad + 2ppre + ppee$

$- pprr \quad - ccde$

$+ ccdr$

$f[\varepsilon.\mathrm{I}])$ *On solving equations by reduction*

The equations to be reduced: $xxxz = -ppaa + 4raaa + aaaa$

$$xxxz = -ppaa - 4raaa + aaaa$$
$$xxxz = +ppaa - 4raaa - aaaa$$
$$xxxz = +ppaa + 4raaa - aaaa$$

$aaaa \pm 4raaa - ppaa = + xxxz$

$aaaa \pm 4raaa = + xxxz + ppaa$

add $+ 6rraa \pm 4rrra + rrrr$ [to each side]

$aaaa \pm 4raaa + 6rraa \pm 4rrra + rrrr = + xxxz \ \pm 4rrra + ppaa$
$$+ rrrr \qquad\qquad + 6rraa$$

I^{st} Let $\pm a + r = e$

$$a = \pm e \mp r$$
$$aa = + ee - 2re + rr$$

$\pm 4rrra = + 4rrre - 4rrrr$

$+ ppaa + 6rraa = + ppee - 2ppre + pprr + 6rree - 12rrre + 6rrrr$

$eeee - 6rree + 8rrre \ = \ \pm xxxz$
$$- ppee \ + 2ppre \qquad + 3rrrr$$
$$+ pprr$$

or

$- eeee + 6rree - 8rrre \ = \ \mp xxxz$
$$+ ppee \ - 2ppre \qquad - 3rrrr$$
$$- pprr$$

284

෯

2nd Let $\mp a - r = e$

$$a = \mp e \mp r$$
$$aa = + ee + 2re + rr$$

$$\pm 4rrra = - 4rrre - 4rrrr$$
$$+ ppaa + 6rraa = + ppee + 2ppre + pprr + 6rree + 12rrre + 6rrrr$$

$$\pm xxxz = - 8rrre - 6rree + eeee$$
$$+ 3rrrr \quad - 2ppre - ppee$$
$$+ pprr$$

or

$$\mp xxxz = + 8rrre + 6rree - eeee$$
$$- 3rrrr \quad + 2ppre + ppee$$
$$- pprr$$

Add MS 6783 f. 147

$f[\varepsilon.2])$ *On solving equations by reduction*

The equations to be reduced: $xxxz = - ppaa - 4raaa - aaaa$

$$xxxz = - ppaa + 4raaa - aaaa$$

$\pm a + r = e$ $\mp a - r = e$

$a = \pm e \mp r$ $a = \mp e \mp r$

$$eeee - 6rree + 8rrre = - xxxz = - 8rrre - 6rree + eeee$$
$$+ ppee - 2ppre \quad + 3rrrr \quad + 2ppre + ppee$$
$$- pprr$$

or

$$- eeee + 6rree - 8rrre = + xxxz = + 8rrre + 6rree - eeee$$
$$- ppee + 2ppre \quad - 3rrrr \quad - 2ppre - ppee$$
$$+ pprr$$

$f[\varepsilon.3])$ *On solving equations by reduction*

The equations to be reduced:

$$xxxz = -ccda + 4raaa + aaaa$$
$$xxxz = +ccda - 4raaa + aaaa$$
$$xxxz = +ccda - 4raaa - aaaa$$
$$xxxz = -ccda + 4raaa - aaaa$$

$aaaa \pm 4raaa \mp ccda = +xxxz$

$aaaa \pm 4raaa = +xxxz \pm ccda$

add $+6rraa \pm 4rrra + rrrr$ [to each side]

$aaaa \pm 4raaa + 6rraa \pm 4rrra + rrrr = +xxxz \pm ccda + 6rraa$
$$+ rrrr \pm 4rrra$$

1st Let $\pm a + r = e$

$$a = \pm e \mp r$$
$$aa = +ee - 2re + rr$$

$\pm ccda \pm 4rrra = +ccde - ccdr + 4rrre - 4rrrr$

$+6rraa = +6rree - 12rrre + 6rrrr$

$eeee - 6rree + 8rrre = \pm xxxz$
$$- ccde \qquad\quad + 3rrrr$$
$$- ccdr$$

or

$-eeee + 6rree - 8rrre = \mp xxxz$
$$+ ccde \qquad\quad - 3rrrr$$
$$+ ccdr$$

286

2nd Let $\quad \mp a - r = e$

$$a = \mp e \mp r$$

$$aa = + ee + 2re + rr$$

$$\pm ccda \pm 4rrra = - ccde - ccdr - 4rrre - 4rrrr$$
$$+ 6rraa = + 6rree + 12rrre + 6rrrr$$

$+ xxxz = - 8rrre - 6rree + eeee$

$+ 3rrrr \quad + ccde$

$- ccdr$

or

$- xxxz = + 8rrre + 6rree - eeee$

$- 3rrrr \quad - ccde$

$+ ccdr$

$f[\varepsilon.4])$ *On solving equations by reduction*

The equations to be reduced: $xxxz = - ccda - 4raaa - aaaa$

$\qquad\qquad\qquad\qquad\qquad\qquad xxxz = + ccda + 4raaa - aaaa$

$\pm a + r = e$ $\mp a - r = e$

$a = \pm e \mp r$ $a = \mp e \mp r$

$eeee - 6rree + 8rrre = - xxxz = - 8rrre - 6rree + eeee$

$\qquad\qquad + ccde \qquad + 3rrrr \qquad - ccde$

$\qquad\qquad\qquad\qquad\quad + ccdr$

or

$- eeee + 6rree - 8rrre = + xxxz = + 8rrre + 6rree - eeee$

$\qquad\qquad\quad - ccde \qquad - 3rrrr \qquad + ccde$

$\qquad\qquad\qquad\qquad\quad - ccdr$

Appendix

Correlations between Harriot's manuscripts and the texts of Viète, Warner and Torporley

Operations of arithmetic in letters

Torporley's description
'The exemplified forms of the four operations of arithmetic in letters.'[1]

Manuscripts
Sheets 1) to 4)[2]   British Library Add MS 6784, ff. 322–325.
Torporley's *Summary*   Sion College Arc MS L.40.2/L.40, ff. 35–35ᵛ.

Correlation between *Operations of arithmetic in letters*, the *Isagoge* and the *Praxis*:

Operations of arithmetic		*Isagoge*	*Praxis*
Sheet 1)	Addition, subtraction		7–8
Sheet 2)	Multiplication, division		8–9
Sheet 3)	Fractions	Chapter IV, Precept IV	10
Sheet 4)	Equations	Chapter IV, Precept V	11

Treatise on equations

Section (a): On solving equations in numbers

Torporley's description

> 'In the paragraph supposed *a*)and in 13 sheets are three examples of
> quadratics, of which the first is [Harriot's], the other two are Viète's,
> and five cubics, all Viète's apart from the first. And five quartics of
> which the fourth is [Harriot's], the rest Viète's. And these, according
> to the method of Viète are all affirmative equations.'[3]

There are actually 12 sheets: Torporley originally wrote '12' but later changed
this to '13'. The first *and* third quadratics are Harriot's own examples; so is
the *third* quartic. Torporley's description is otherwise correct.

Manuscripts

Sheets 1) to 12)	Add MSS 6782, ff. 388–399 (in reverse order).
Torporley's *Summary*	Sion College Arc MS L.40.2/L.40, ff. 49^v.

Correlation between Section (a) and Viète's *De resolutione potestatum resolutione*:

Treatise on equations Section (*a*)		*De numerosa potestatum resolutione* Problems 1 to 6	
a.1)	$aa + 24a = 2356$		
a.2)	$aa + 7a = 60750$	1	1 Q + 7N equals 60750
a.3)	$aa + 762a = 22120$		
a.4)	$aaa + 35a = 2932$		
a.5)	$aaa + 30a = 14356197$	2	1C + 30N equals 14356197
a.6)	$aaa + 95400a = 1819459$		95400N + 1C equals 1819459
a.7)	$aaa + 30aa = 86220288$	3	1C + 30Q equals 86220288
a.8)	$aaa + 10000aa = 5773824$		10000Q + 1C equals 5773824
a.9)	$aaaa + 10000a = 355776$	4	1QQ + 1000N equals 355776
a.10)	$aaaa + 100000a = 2731776$		100000N + 1QQ equals 2731776
	$aaaa + 300aaa = 4478976$		
a.11)	$aaaa + 10aaa = 470016$	5	1QQ + 1C equals 470016
a.12)	$aaaa + 200aa + 100a$ $= 449376$	6	1QQ + 200Q + 100N equals 449376

Section (b): On solving equations in numbers

Torporley's description

'The other part of it, as paragraph *b*) in 12 sheets has, as Viète has, analysis of powers negatively affected: quadratics in *b*1), *b*2), *b*3), cubics in *b*4) to *b*10), quartics in *b*10), *b*11), *b*12).'[4]

Manuscripts

Section (*b*) is the only section of Harriot's treatise now remaining at Petworth.

Sheets *b*.1) to *b*.12) Petworth MS HMC 241.1, ff. 1–9, 11–13 (reversed)
Torporley's *Summary* Sion College Arc MS L.40.2/L.40, ff. 50–50ᵛ.

Correlation between Section (b) and Viète's *De numerosa potestatum resolutione*:

Treatise on equations	*De numerosa potestatum resolutione*
Section (*b*)	Problems 10 to 15

b.1)	$aa - 7a = 60750$	10	$1Q - 7N$ equals 60750
b.2)	$aa - 240a = 484954$		$1Q - 240N$ equals 484954
b.3)	$aa - 60a = 1600$		$1Q - 60N$ equals 1600
	$aa + 8a = 128$		$1Q + 8N$ equals 128
b.4)	$aaa - 10a = 13584$	11	$1C - 10N$ equals 13584
b.5)	$aaa - 116620a = 352947$		$1C - 116620N$ equals 352947
b.6)	$aaa - 6400a = 153000$		$1C - 6400N$ equals 153000
	$aaa + 64a = 1024$		$1C + 64N$ equals 1024
b.7)	$aaa - 7aa = 14580$	12	$1C - 7Q$ equals 14580
b.8)	$aaa - 10a = 288$		$1C - 10N$ equals 288
b.9)	$aaa - 7aa = 720$		$1C - 7Q$ equals 720
	$aaa + 8aa = 1024$		$1C + 8Q$ equals 1024
b.10)	$aaaa - 68aaa + 202752a$ $= 5308416$	13	$1QQ - 68C + 202752N$ equals 5308416
	$100000a + aaaa = 2731776$		$100000N + 1QQ$ equals 2731776
b.11)	$aaaa + 10aaa - 200a$ $= 1369856$	14	$1QQ - 10C + 200N$ equals 1369856
b.12)	$aaaaa - 5aaa + 500a$ $= 7905504$	15	$1QC - 5C + 500N$ equals 7905504

Section (c): On solving equations in numbers

Torporley's description

'The third part of this, as paragraph *c*), has 18 sheets, and treats the analysis of avulsed powers, as Viète, where there are multiple roots, and the limits of each are demonstrated. The examples of this are two quadratics, four cubics, two quartics.'[5]

Section (*c*) actually discusses only one quadratic, two cubics and two quartics.

Manuscripts

Sheets *c*.1) to *c*.18)   British Library Add MS 6782, ff. 400–417 (reversed)
Torporley's *Summary*   Sion College Arc MS L.40.2/L.40, ff. 50^v–51^v.

Correlation between Section (c) and Viète's *De numerosa potestatum resolutione*:

Treatise of equations		*De numerosa potestatum resolutione*	
Section (*c*)		Problems 16 to 20	
c.1) to *c*.2)	$370a - aa = 9261$	16	$370N - 1Q$ equals 9261
c.3) to *c*.6)	$13104a - aaa = 155520$	17	$13104N - 1C$ equals 155520
c.7) to *c*.10)	$57aa - aaa = 243000$	18	$57Q - 1C$ equals 243000
c.11) to *c*.13)	$27755a - aaaa$	19	$27755N - 1QQ$ equals
	$= 217944$		217944
c.14) to *c*.18)	$65aaa - aaaa$	20	$65C - 1QQ$ equals 1481544
	$= 1481544$		

Section (d): On the generation of canonical equations

Torporley's description

> 'The first part thus: *On the generation of canonical equations*, 21 sheets
> on that theme, paginated together under paragraph *d*) with two
> appendices on the multiplication of roots.'[6]

Manuscripts

Sheets *d*.1) to *d*.21)	British Library Add MS 6783, ff. 163–183 (reversed).
Sheet *d*.7.2)	British Library Add MS 6783, f. 204.
Sheet *d*.13.2)	British Library Add MS 6783, f. 156.
Torporley's *Summary*	Sion College Arc MS L.40.2/L.40, ff. 41^v–43^v, 47^v.

Correlation between Harriot's Section (d) and the *Praxis:*

Section (*d*)	*Praxis*		
	Section 2 Propositions	Section 3 Problems	Section 4 Propositions
d.1)	2, 1	1	1, 2
d.2)	5		5
d.3)	4	2, 3	4, 6
d.4)	3	4	3, 7
d.5)		5	9
d.6)	8, 6, 7		20, 18, 19
d.7)	12, 11	9–11	24, 22, 25–27
d.7.2)	–	–	–
d.8)	9	12–14	21, 28–30
d.9)	10	15–17	23, 31–33
d.10)		18, 19	34, 35
d.11)		20	36
d.12)		21	37
d.13)	15, 14, 13		40, 39, 38
d.13.2)	–	–	–

The remaining sheets of Section (*d*) are represented in the *Praxis* as follows:

d.14) to *d*.16)	Section 2, pages 46–51, reordered
d.17) to *d*.19)	–
d.20) to *d*.21)	Section 6, pages 87, 88

Section (e): On solving equations by reduction

Torporley's description

> 'The second part moreover under the title: *On solving equations by reduction* has 29 sheets as paragraph *e*).'[7]

Manuscripts

Sheets *e*.1) to *e*.14)   British Library Add MS 6783, ff. 98–112.
Sheets *e*.15 to *e*.29)   British Library Add MS 6783, ff. 184–198 (reversed).
Sheets *e*.3.2) to *e*.3.6)   British Library Add MS 6783, ff. 215–219.
Torporley's *Summary*   Sion College Arc MS L.40.2/L.40, ff. 44–46ᵛ.

(i) Correlation between Section (e) and material to be found in Viète:

Section (*e*)	Viète's *De aequationum recognitione*, Chapter II (equations containing cube and square terms only)
e.15)	Problem II
e.16)	Problem I
e.17)	Problem III

Section (*e*)	Viète's *De aequationum recognitione*, Chapter II (equations containing cubes, squares and linear terms)
e.18)	Problem I
e.19)	Problem VII
e.20)	Problem II
e.21)	Problem VI
e.22)	Problem IV
e.23)	Problem V
e.24)	Problem III

Section (*f*)	Viète's *De numerosa potestatum resolutione*
e.26) and *e*.27)	following Problem XVIII

(ii) Correlation between Section (e) and the *Praxis*:

Section (*e*)	*Praxis*
e.1) to *e*.4)	–
e.5)	Section 6, Problem 12
e.6)	Section 6, Problem 13
	Section 5, Proposition 1 (statement only)
e.7)	Section 5, Proposition 1 (statement only)
e.8)	Section 5, Lemmas 1–4
e.9)	Section 5, Lemma 5
e.10)	Section 5, Proposition 2 (statement only)
e.11)	Section 5, Proposition 3 (statement only)
e.12)	Section 5, Proposition 4 (statement only)
e.13) to *e*.14)	–
e.15) to *e*.25)	Section 6, Problems 1–11, in the same order
e.26) to *e*.27)	–
e.28)	Section 5, Lemma 6
e.29)	Section 5, Lemma 7

Section (f): On solving equations by reduction

Torporley's description

'Under the same heading [as (*e*)], *f* α) 7 sheets; *f* β) also 7 sheets; and following these in the numeration of the sheets, *f* γ) to sheet *f* 18 γ) with an appendix with two lemmas that should not be disparaged, omitted by them, then *f* δ) 8 sheets; *f* ε) 4 sheets; *f* ζ) also 4 sheets.'[8]

Manuscripts

f α (7 sheets)	British Library Add MS 6783, ff. 113–118, 130.
f β (7 sheets)	British Library Add MS 6783, ff. 139–145 (reversed).
f γ (11 sheets)	British Library Add MS 6783, ff. 150–155, 157–160, 162.
f δ (8 sheets)	British Library Add MS 6783, ff. 131–138 (reversed).
f ε (4 sheets)	British Library Add MS 6783, ff. 146–149.
f ζ (4 sheets)	British Library Add MS 6783, ff. 199–202.
Torporley's Summary	Sion College Arc MS L.40.2/L.40, ff. 46ᵛ–49.

The contents of parts α to ζ are as follows:

f α), *f* β)	quartics lacking square or linear terms, or having all four terms
f γ)	quartics with all four terms
f δ)	quartics lacking the cube term
f ε)	quartics with all four terms
f ζ)	quartics lacking the square or linear term

Parts α and β contain almost identical material (apart from the numerical examples). Part α appears to be the later and more finished version and I have therefore omitted part β from this edition.

The pagination of γ runs straight on from that of α (or, equivalently, β). In both α and γ, Harriot usually dealt with only one equation at a time, but sometimes with two simultaneously. In part ε, however, he explained how it was possible to handle four equations at a time, and in ζ he handled four equations at a time from the start. Parts ε and ζ therefore appear to have been written out later than α and γ.

I have chosen the order δ, α, γ, ζ, ε, because then Section (*f*) treats quartic equations in the same way that Section (*e*) treats cubics: it opens with the general method for quartics lacking the cube term and then goes on to show how to remove the cube term from any other quartic. This ordering also corresponds roughly with that found in Section 6 of the *Praxis*.

Correlation between Section (f) and the *Praxis*

Section 6 of the *Praxis* contains numerous examples of removing the cube term from a quartic. These examples are taken from parts α, γ, ζ and ε as shown below. The material in part δ, however, is not represented in the *Praxis* at all (so the purpose of removing the cube term is never made clear).

Manuscript sheet	*Praxis*, Section 6
$f\delta$ 1) – 8)	–
	Problems
$f\alpha$ 1)	14, 15
$f\alpha$ 2)	16, 17
$f\alpha$ 3)	18, 19
$f\alpha$ 4), 5)	20, 21
$f\alpha$ 6), 7)	22, 23
$f\gamma$ 8)	24
$f\gamma$ 9) –16)	–
$f\gamma$ 17)	27
$f\gamma$ 18)	–
$f\zeta$ 1) (part)	24
$f\zeta$ 2) (part)	25, 30
$f\zeta$ 3) (part)	26, 27
$f\zeta$ 4) (part)	28, 29
$f\varepsilon$ 1) (part)	31, 32
$f\varepsilon$ 2)	–
$f\varepsilon$ 3) (part)	33, 34
$f\varepsilon$ 4)	–

Notes

Introduction

1. For biographies of Harriot see Stevens 1900; Shirley 1983.

2. Most of Harriot's surviving manuscripts are now held in the British Library, as Add MSS 6782–6789; copies of the British Library papers are held in the University Libraries of Cambridge, Durham and Oxford. A smaller selection of papers remains at Petworth House, as MSS HMC 240–241; copies of the Petworth papers are held in the West Sussex Record Office at Chichester. For the history of the papers and attempts at publication see Shirley 1983, 1–38; for a survey of the scientific and mathematical contents see Lohne 1979.

3. For an account of this and related voyages see Milton 2000.

4. Harvey 1593, 190; Hues 1594, 166. For other contemporary references to Harriot see Quinn and Shirley 1969.

5. For biographies of Harriot's colleagues and companions see Shirley 1983, 358–379, 388–424.

6. Add MS 6788, ff. 117–117^V, reproduced in Pepper 1967a, 290. Pepper dates the letter to 1586, but Viète was dismissed from the court in Paris in 1584 and lived in or near Tours for the next ten years. He returned to Paris only in 1594, and only from that year were his books printed there. It seems more plausible, therefore, that Torporley's letter was written some time after 1594.

7. John Pell to an unknown recipient 12 October 1642, Halliwell 1841, xv; Aubrey 1898, II, 263.

8. Add MS 6782, ff. 482–483.

9. '*Si B planum in A minus A cubo, aequetur Z solido: est B planum compositum ex quadratis trium proportionalium: & Z solidum quod fit ductu alterius extrema in aggregatum quadratum a reliquis: & fit A prima vel tertia. Sunto proportionales 2, $\sqrt{20}$, 10 dicetur 124N − 1C aequalis 240 et fit 1N 2 vel 10. *'; Viète 1646, III, Theorema II.

10. The *Arithmetic* of Diophantus was first published in a translation by Guilielmus Xylander, Basel 1575; Books I to IV of the *Conics* of Apollonius were published first by Joannes Baptista Memus, Venice 1537, then by Federico Commandino, Bologna 1566; the *Collections* of Pappus of Alexandria were also published by Commandino, Pisauri 1588.

11. Bombelli 1572; Viète 1593a; see Reich 1968.

12. For Viète, *poristic* was the method of testing or verifying a theorem while *zetetic* was the art of setting up equations by which an unknown quantity could be found. To these he added *exegetic*, the process of solving equations.

13. 'Denique fastuosum problema problematum ars Analytice, . . . jure sibi adrogat, Quod est, NULLUM NON PROBLEMA SOLVERE'; Viète 1646, 12, paragraph 29. Viète promised nine further books to demonstrate the use and application of the algebraic method in arithmetic, geometry and trigonometry. These books together were to constitute the *Opus restitutae mathematicae analyseos, seu algebra nova*. Seven of them were published individually over the next few years, and were collected together in Viète's *Opera mathematica* edited by Frans Van Schooten in 1646. They are now available in English translation in Witmer 1983.

14. *De aequationum recognitione* was published in Paris in 1615, edited by Alexander Andersen, but was mentioned by name in the *Supplementum geometriae* published in 1593: 'Enimvero ostensum est in tractatu de aequationum recogntione, aequationes quadrato-quadratorum ad aequationes cuborum reduci.'; Viète 1646, 257.

15. For the oriental origins of the algorithms expounded by Viète see Rashed 1974; Chemla 1994.

16. *De potestatum resolutione* was published with Viète's consent by Marino Ghetaldi in Paris in 1600, but the power of the methods contained in it was already hinted at in the *Isagoge* in 1591: *Potestatum porro quarumcumque, sive purarum sive (quod nesciverunt veteres atque novi) adfectarum, tradit Ars resolutionem*'; 'Next the Art treats the resolution of powers of any kind, whether pure or (which neither the ancient or modern writers knew) affected'; Viète 1646, 12, paragraph 23.

17. On the first sheet of the *Treatise on equations* (Add MS 6782, f. 399) Harriot wrote 'Vieta. fol. 7. b'. In the 1600 edition of Viète's *De potestatum resolutione*, f. 7^v was where Viète began his treatment of affected equations.

18. The original of this letter is now lost but it is reproduced in Stevens 1900, 121–122 and Shirley 1983, 1–2. Ghetaldi's *Promotus Archimedes, seu de variis corporum generibus gravitate et magniutdine comparatis* was published in 1603, and Viète's *De potestatum resolutione* in 1600.

19. Stifel 1544, 252. For a single unknown, Stifel used cossist notation in which R represented the unknown quantity, Q or Z its square and C its cube. This led to the use of Z to represent a quantity in its own right, as in '3Z times 4B gives 12ZB' but also to indicate the operation of squaring as in '1A squared gives 1AZ' (both examples from Stifel 1544, 252).

20. Bombelli 1572; Stevin 1585. A similar system of notation had been introduced by Nicholas Chuquet in 1484 who wrote, for example, 6^3 for $6x^3$, but Chuquet's notation was changed when Etienne de la Roche published some of his work in 1520 and was never taken up by others; see Flegg, Hay and Moss 1985.

21. Recorde 1557; see also Cajori 1928–29, 297–298.

22. In Torporley's manuscripts, written c.1630–1632, the inequality signs already appear in their modern form < and > without the cross-strokes.

23. The modern division sign was not introduced until later in the seventeenth century, in Pell and Rahn 1668.

24. Cajori 1928–29, 229–250.

25. The first appearance of the ± sign in print was in Oughtred 1631; see Cajori 1928–29, 245.

26. The treatment of separate cases according to the signs of the coefficients went back to al-Khwārizmī's seminal text, his *Al-jabr wa'l-muqābala* (*c.* 825) and was followed in every European text until well into the seventeenth century.

27. *Operationes logisticae in notis.*

28. Add MS 6784, f. 324; Viète 1646, 7–8; Witmer 1983, 23.

29. '*Oporteat A plano/B addere Z quadratum/G. Summa erit G in A planum + Z quadratum/ B in G*'; Viète 1646, 8.

30. Add MS 6784, f. 324.

31. '*Vel, B in G oporteat adplicere ad A planum/D. Ducta utraque magnitudine in D, ortiva erit B in G in D/A plano*'; Viète 1646, 8.

32. Add MS 6784, f. 324.

33. Instructions on (i) changing terms from one side to another and (ii) reducing the leading coefficient to one were given by al-Khwārizmi *in Al-jabr w'al-muqābala* (*c.* 825) and were repeated in many sixteenth-century texts, for example, Peletier 1554; Mennher 1556; Recorde 1557; Nuñez 1564. Advice on (iii) division by excess powers of the unknown can be found in, for example, Recorde 1557; the earliest instance I have found of it given as an explicit rule is in Perez de Moya 1573. Rules (i) to (iii) were the first of the ten rules for reducing equations given by Simon Stevin in Stevin 1585, 63–65.

34. Viète 1646, 9; Witmer 1983, 25–27.

35. '*Proponatur A quadratum minus D plano aequari G quadrato minus B in A. Dico A quadratum plus B in A aequari G quadrato plus D plano, neque per istam transpositionem sub contraria adfectionis nota aequalitatem immutari*'; Viète 1646, 9.

36. Add MS 6784, f. 325.

37. It is possible that Viète had also begun to think of polynomials as products of factors. At the end of *De aequationum recognitione* he noted that the coefficients of equations with 2, 3, 4 or 5 positive roots are what are now called symmetric functions of those roots. He closed by saying that he had otherwise treated at some length the reasoning behind this elegant and beautiful observation, the end and Crown of his work: '*Atque haec elegans et perpulchrae speculationis sylloge, tractatui alioquin effuso, finem aliquem et Coronida tandem imponito*'; Viète 1646, 158; Witmer 1983, 309–310. Nothing else on the subject is to be found in Viète's published work.

38. Harriot did not attempt to prove that all polynomials can be written as a product of factors, nor did he pose the question since he was interested only in a restricted class of polynomial equations, those having one or more positive real roots.

39. John Wallis in his account of Harriot's algebra pointed out that the positive roots of an equation together with the positive roots of its conjugate, establish all the real roots, Wallis 1685, 158–159.

40. Cajori 1928; Pycior 1997, 54–64; especially, 57, 64.

41. The discriminant of the equation $x^3 - 3p^2x = 2q^3$ is $q^6 - p^6$, which can be used to discover the number of distinct real roots. In Harriot's equations, b and c are assumed positive, so the conditions $c > b$, $c = b$, $c < b$ correspond to $q^6 - p^6 > 0$, $q^6 - p^6 = 0$, $q^6 - p^6 < 0$.

42. Cardano 1545, Cap XXXIX; Bombelli 1572, 353–411; Stevin 1585, 81–87; Viète 1646, 140–148.

43. For full transcripts of Harriot's Will see Stevens 1900, 193–203; Tanner 1967b, 244–247.

44. Shirley 1983, 413–414.

45. British Library Add MS 6789, ff. 448–450; reprinted in Tanner 1969, 346–349.

46. The *Congestor* survives in two manuscript copies. The first was held for many years at Sion College in London, where Torporley spent his final years, but was transferred along with all other Sion College manuscripts to Lambeth Palace Library in 1996. The Sion College MS catalogue listed it under the title *Congestor analiticus*; Anthony Wood in his account of Torporley referred to it as *Congestor opus mathematicum*; Wood 1691–92, II, 525. Its modern shelfmark is Sion College MS Arc L.40.2/L.40, ff. 1–34^V. The second copy is in the Macclesfield collection, acquired by Cambridge University Library in 2001. For a detailed account of the history and contents of the *Congestor* see Tanner 1977.

47. *Congestor*, f. 5.

48. Identification of prime numbers, *Congestor* ff. 5–25. 'Thomas Hariotus examinatur Stifelius de numeris diagonalibus', *Congestor*, ff. 26–34^V; see Tanner 1977, 419–428.

49. ' ... *me licet hostis inter alia convitia et hoc criminaretur domino Petworthiae quod essem dialecticus ignarus*', Sion College MS Arc L.40.2/L.40, f. 11; Halliwell 1841, 114.

50. '... *hominis per eos in coelum sublati* ...', Sion College MS Arc L.40.2/L.40, f. 8; Halliwell 1841, 110.

51. When the *Praxis* was published in 1631 no editor's name appeared on the title page, perhaps because the Executors knew that they had contravened the terms of Harriot's Will. A letter from Aylesbury to the Earl of Northumberland makes it clear, however, that it was Warner who did most of the work and saw the book through the press (Aylesbury to Percy, 5 April 1632, Add MS 4396, f. 90; Halliwell 1841, 71). Warner's role was also well known to his friends, Charles Cavendish, Robert Payne and John Pell. For a modern assessment of Warner and his work see Prins 1992.

52. '*Definitio 14*
    Est etiam aliud quoddam aequationum genus; quae licet canonicae non sint, cum tamen ab iis, ut ab originalibus suis aequationes canonicae deriventur, canonicarum originales in sequentibus denominantur.'; *Praxis*, 4.

53. '*Definitio 15*
    Primario canonicarum species est earum quae ab originalibus per derivationem constituuntur.'; *Praxis*, 4.

54. '*Definitio 16*
    Secundaria canonicarum est earum quae a primariis per reductionem constituuntur.'; *Praxis*, 5.

55. *Praxis*, 12–15.

56. *Praxis*, 16–26.

57. *Praxis*, 29–46.

58. *Praxis*, 52–77.

59. Sion College MS Arc L.40.2/E.10, ff. 7–12. The Latin text, not always accurately transcribed, can be found in Halliwell 1841, 109–116; a preliminary and partial translation is given in Stedall 2000a, 471–473.

60. '... *non possum non conqueri, illud ne male habere quod ejus glossarii ita penitus transformarunt eandem, ut non solum non ordinem sed vix verbum ejus retineant. Id quod forte laude dignum esset, si alicujus illiterati fortuito inventa distribuissent*'; *Corrector*, f. 8; Halliwell 1841, 110.

61. '*Operationes logisticae in notis ita dictae ab Harrioto, ubi incipiunt ejus glossatores sub titulo. Logisticae speciosae quatuor operationum formae exemplificatae non ita scilicet magistraliter distanti illo.*' *Corrector*, f. 9; Halliwell 1841, 111.

62. '*Primo accurata tractatio irrationalis surdorum sive, ut ille vocat eos, radicalium numerorum, non illa quidem aliena ab analytica arte confitentibus ejus interpretibus in sectione, sed totam artem omittentibus mixum. Nam si inutilis ad Exegesin, cur ejus ibi fecerunt mentionem? Si mentionem fecerunt, certe ut non inutilis, cur igitur non descipserunt?*' *Corrector*, f. 8; Halliwell 1841, 110. Radicals are mentioned in the *Praxis*, 99–101.

63. '*In ipso analyticae artificio contentus trimembri divisione inscribit primam ejus partem ita. De generatione aequationum canonicarum sub paragrapho d) compaginatis ad illud argumentum chartis 21 cum appendiculis duobus de multiplicatione radicum.*

Secunda pars autem sub titulo De resolutione per reductionem, habet paragraphum e) chartas 29. Item ∫ α) chartas 7: ∫ β) chartas quoque 7: et succedens illis in chartarum numeratione, ∫ γ) ad chartam ∫ 18 γ) cum appendicula sub lemmata duplici non illa contemnenda licet a suis omissa: Deinde ∫ δ) chartae 8. ∫ ε) chartae 4. ∫ ζ) item 4: Postremo chartae novem continentes reductiones veterum ad Harrioti methodum revocatas.

Sed tertiam partem (non ita studio dissentiendi) cum Vieta inscribit. De numerosa potestatum resolutione, et recte merito. Non totus fere est Vietaeus per exempla singula. Et supposito paragrapho a) et in chartis 13 sunt exempla tria quadratica quorum primum est suum, duo reliqua sunt Vietae, quinque cubicae omnia Vietae praeter primum. Et quinque quadrato quadratica quorum quartum est suum, reliqua Vietae. Et sunt ista secundum Vietae methodum aequationum omnino affirmantium. Altera ejus pars sub paragrapho b) in chartis 12 habet cum Vieta habet analyticam potestatum affectarum negate quadratica b 1) b 2) b 3) cubica b 4) ad b 10) quadrato-quadratica b 10) b 11) b 12).

Tertia ejus pars sub paragrapho c) habet 18 chartas, tractat analysin potestatum avulsurum cum Vieta, ubi radices sunt multiplices et singularum limites demonstrantur. Exempla hujus sunt quadratica duo, cubico-cubica 4, quadrato-quadratica duo.'; *Corrector*, ff. 9–9^v; Halliwell 1841, 111.

64. Halliwell wrongly has at this point '*cum Vieta in suo libro*'.

65. In Halliwell's transcript, the letter *a* is missing, but it is clear in the original. Harriot himself did not give a letter to his first section, but Torporley naturally adopted the letter *a* in keeping with Harriot's usage in subsequent sections.

66. The second subdivision of Part three.

67. The third subdivision of Part three.

68. 'Avulsed powers' is the term used by Viète for equations of the form $ba^m - a^n = 0$ when $m < n$.

69. In the *Praxis* too, the theoretical treatment of equations precedes the section on numerical solution.

70. For a detailed comparison of Torporley's description in the *Corrector* with the surviving manuscripts, see the Appendix.

71. Sion College MS Arc L.40.2/L.40, ff. 35–54^V. Torporley's *Summary* has received even less attention than his *Corrector*. Stevens 1900, 170, briefly described it as comprising 'examples of Algebraic processes'. Seaton 1956 mentioned it in a non-mathematical context. Tanner 1974, 100–101, was the first to recognize its importance: 'Torporley has meticulously crammed symbol for symbol well over 150 large sized pages of Harriot's mathematical writings into the twenty folios (35 to 55)'; see also further references in Tanner 1980, 137, 148.

72. Tanner 1974, 100, wrote: 'It is, however, astonishing and gratifying for us who are thus enabled to gather together in correct sequence correlated pages with relatively little further toil.' In note 48, Tanner made a preliminary list of the relevant manuscript sheets.

73. 'Problems 16 and 17, with changed signs, in Warner, and 18'; 'omitted in Warner'; *Summary*, f. 42^V. Problems 16, 17 and 18 referred to here occur in Section 3 of the *Praxis*, 43–45.

74. *Summary*, ff. 35^V–41.

75. Aylesbury to Percy, 5 April 1632, Add MS 4396, f. 90.

76. Hartlib papers 1639, 30/4/9B.

77. Cavendish to Pell, 6 October 1651, Add MS 4278, f. 321; printed in Hervey 1952, 89.

78. The surviving mathematical papers of Charles Cavendish are preserved in the British Library as Harley MSS 6001–6002, 6083. Photocopies are kept alongside the Harriot manuscripts in the University Libraries of Cambridge, Durham and Oxford. Cavendish's copy of Harriot's *De numeris triangularibus* is in Add MS 6083, ff. 403–455 and contains references to 'Mr. Hariot's loose papers', in ff. 403^V, 404, 429^V. For further references to Harriot's papers see also Add MS 6002, ff. 4, 44.

79. An anonymous narrator later wrote: 'coming to London, [Aylesbury] found his Library, wherein were many rare and curious books, plundered'; Bodleian Library MS Rawlinson B.158, f. 153.

80. Harriot's *De numeris triangularibus* is in Add MS 6782, ff. 107–146; Sections (*a*) and (*c*) of the *Treatise on equations* are in Add MS 6782, ff. 388–417. At the end of the *Summary* Torporley wrote out the full title of *De numeris triangularibus* (*De numeris triangularibus et inde de progressionibus arithmeticis magisteria magna T. H.*, Sion College MS Arc L.40.2/L.40, f. 54^V), which suggests that he may have held a copy of that treatise, as well as of the *Treatise on equations*.

81. From the 1660s onwards Harriot's papers were thought to be lost. The Royal Society instigated searches on more than one occasion, but without success, see Shirley 1983, 7–9. By that time the original copy of Harriot's Will was lost and the only surviving Executor was Robert Sidney, Viscount Lisle (1595–1677), son-in-law of Henry Percy, but he seems to have had little or nothing to do with Harriot's papers after the 1620s.

82. In the late 1620s Cavendish persuaded Oughtred to write his *Clavis mathematicae*, the first book to introduce Viète's notation and some simple algebraic geometry to English readers. The *Clavis* was published in 1631 under the title *Arithmeticae in numeris et speciebus institutio: quae tum logisticae, tum analyticae, atque totius mathematicae quasi clavis est*, London 1631, and the influence of Cavendish was acknowledged in the preface; see Stedall 2000b. Possibly a few years later, Oughtred annotated a copy of Viète's *De aequationum recognitione* owned by Cavendish; the annotations were later copied by John Pell, and can be found in Add MS 4423, ff. 146–153^V.

83. In 1655 Oughtred wrote to John Wallis: '... full twenty years ago, the learned patron of sciences, Sir Charles Cavendish, shewed me a written paper sent out of France, in which were some very few excellent new theorems, wrought by the way, as I suppose, of Cavalieri ...', Oughtred to Wallis, 17 August 1655, Rigaud 1841, I, 87–88. Among Cavendish's papers is a hand-written treatise *Elemens des indivisibles*, Harley MS 6083, ff. 279–302, and it is tempting to conclude that this was the same 'written paper' that he once showed to Oughtred.

84. Add MS 4425, f.11. See also Derand to Cavendish, 11 February 1635, Rigaud 1841, I, 23.

85. Gaukroger 1995, 331.

86. Descartes on Viète: '*Et ainsis i'ay commencé où il avoit achevé; ce que i'ay fait toutesfois sans y penser, car i'ay plus feüilleté Viete depuis que i'ay receu vostre derniere, que ie n'avois fait auparavant, l'ayant trouvé icy par hazard entre les mains d'un de mes amis*'; Descartes to Mersenne, [December 1637?], in Descartes 1897–1910, I, 479–480.

Descartes on Harriot: '... *c'est Henriotti, que ie pensois que Gillot est emporté avec lui, ... l'avois eu desir de voir ce livre, a cause qu'on m'avoit dit qu'il contenoit un calcul pour la geometrie, qui estoit fort semblable au mien*'; Descartes [to Huygens], [December 1638], Descartes 1897–1910, II, 456.

For Viète's influence on Descartes see Bos 2001.

87. The convention of using xx for x-squared remained in use on the continent for much of the seventeenth century and was retained by Newton, who was the first English writer to adopt Descartes' x, y, z for unknown quantities.

88. *Praxis*, 114; Descartes 1637, Book III, 73.

89. Wallis 1685, 198.

90. See, for example Moore 1650; Brouncker *et al.* 1658; Brouncker 1668; Dary 1664; Pell and Rahn 1668; Kersey 1673–74.

91. Gibson 1655, Preface.

92. Collins to Wallis, February 1668, Rigaud 1841, II, 484.

93. Collins to Gregory, 25 March 1671, Rigaud 1841, II, 219. Gibson died in 1657 or 1658 and so it seems that the sentence describing the loan of the papers was incorrectly transcribed by Rigaud and should read: '...were lent to one Gibson, deceased, in anno 1650'.

94. 'Mr. Ward the Prof of Astron is to set out the mathematical and other workes of Warner conc[erning] coyne etc.'; Hartlib papers 1650, 28/1/62A.

95. Thorndike to Pell, 23 December 1652, Add MS 4279, f. 275, printed in Halliwell 1841, 94; conversation between Pell and Thorndike, 17 January 1653, recorded by Pell, Add MS 4279, ff. 276–276^v.

96. '[Justinian Isham] hath gotten all the MS. Mathematical of Warner and ... shewed them Mr. Pell'; Hartlib 1653, 28/2/49A. The Warner papers held by Isham are preserved in Northamptonshire Record Office, IL 3422, VI, ff. 1–23.

97. 'An inventorie of the papers of Mr Warner', reprinted in Halliwell 1841, 95. The inventory lists 23 items, most of them on coinage or logarithmic tables. Item 18 is entitled *De resectione spatii*, a topic treated by Harriot and found more than once among Warner's papers. Item 22 is 'A bundle intituled "Mr Protheroe"'. For a detailed analysis of the items see Prins 1992, 25–27.

98. Add MS 4394, f. 392.

99. See, for instance, Add MS 4413, f. 224, Add MS 4415, f. 83.

100. Leibniz wrote about Pell's and Wallis's views on Harriot as part of a review he wrote in 1686 of Wallis's *A treatise of algebra* (1685). '*Quibus usque adeo exacte consentiunt ea, quae habet Cartesius in magna parte libri tertii Geometriae suae, ut ex Harrioto descripta non levis suspicio sit. Certe Pellius et Wallisius id pro certe habere videntur*'; 'to [Harriot's principles] everything that Descartes has in the greater part of the third book of his *Geometria* so far agrees exactly, so that the suspicion that he wrote it from Harriot cannot be taken lightly. Certainly Pell and Wallis seem to hold that view for certain'; Leibniz 1686, 285. Pell's views are not described in *A treatise of algebra*, so Leibniz may have learned of them from Pell himself when he met him in London in 1676.

101. Stedall 2002, chapter 1.

102. '*& ex cujus ore descripsi quod hac de re dixi; eique postquam erat descriptum, ostendi, (examinandum, immutandum, emendandum pro arbitrio suo, siquid alias dictum malit) antequam prelo subjiceretur, totumque illud quod inde prodiit, assentiente & approbante Pellio dictum est.*' 'De Harrioto addenda', following the preface (unpaginated) in Wallis 1693. For Pell's unwillingness to have his name mentioned in print, see Stedall 2002, chapter 5. See also note 100.

103. The copy of the *Praxis* used by Wallis was originally given to Robert Payne by Charles Cavendish, and was later acquired by the Bodleian Library in Oxford; its modern shelfmark is Savile O.9. Wallis's annotations on the flyleaf are printed in Shirley 1983, 10–11.

104. Wallis 1685, 128.

105. Wallis 1685, 129–130.

106. Morland to Wallis, 8 January 1689 and Wallis to Morland, 12 March 1689, Wallis 1693, 209–210.

107. Wallis 1685, 198.

108. Wallis 1685, 126.

109. '*Ce n'est que sur de vaines conjectures ou par un mouvement d'envie que des gens ont voulu faire croire de son vivant meme qu'il avoit tiré sa méthode des autres, & particuliérement d'un certain Harriot Anglois, qu'il n'avoit jamais lu, comme il le déclare dane une de ses Lettres. Et lorsque Monsieur Wallis, un peu trop jaloux de la gloire que la France s'est*

acquise dans les Mathématiques, vient renouveller cette accusation ridicule, on est en droit de ne le point croire, puis qu'il parle sans preuve.' Prestet 1689, II, Preface (unpaginated).

110. '... *la conformité de ses sentimens avec ceux de Harriot touchant la nature des Equations a paru un préjugé raisonnable, pour faire croire qu'il avoit quelque obligation à cet Auteur, quoy qu'il ne l'eut point fait connoitre en public. Celuy qui découvrit le prémier cette conformité fut Mylord Candische, qui se trouva pour lors à Paris, et qui la montra à M. de Roberval. La chose devint ensuite toute publique par le zèle que M. de Roberval faisoit paroitre à diminuer par tout la gloire de M. Descartes. Mais M. Pell, Mathématicien Anglois, le Chevalier Ailesbury, qui avoit été l'exécuteur testamentaire de Harriot et le dépositaire de ses papiers, et meme Guill Warner, qui a fair imprimer son livre, jugeoient plus favourablement de M. Descartes, rejetant tout l'avantage de la conformité sur la personne de Harriot, à qui il étoit assez glorieux que M. Descartes se fut rencontré avec luy.'* Baillet 1691, Livre VIII, 541. The story is not in the abridged English translation, Baillet 1693.

111. See notes 100 and 102.

112. Pell to Cavendish, 2/12 March 1646, Add MS 4280, ff. 117–118; printed in Hervey 1952, 77–79.

113. In his *Idea of mathematicks*, published anonymously in 1638, Pell proposed the setting up of a catalogue of mathematicians and mathematical ideas, and a library of key mathematical texts; see Pell 1638; Descartes 1682, and for translations of both pieces see Fauvel and Gray 1987, 310–314.

114. Collins to Oldenburg for Leibniz, undated, Rigaud 1841, I, 247.

115. Montucla 1799–1802, II, 111, wrote that Wallis's account was inexcusable; Cantor 1894–1908, III, 4, described it as 'nationalist polemic'.

116. Wallis 1685, 128.

117. Hutton 1795–96, I, 91. The same paragraph is in Hutton 1812, II, 286.

118. Hutton 1795–96, I, 586.

119. Rigaud 1833, 52.

120. Shirley 1983, 18–20; Fauvel, Flood and Wilson 2000, 160.

121. Rigaud's notes on Harriot and his work are to be found in Bodleian Library MS Rigaud 9, 35, 51, 56, 61.

122. Rigaud 1833, plate v.

123. Lohne 1979, 305, note *.

124. Tanner 1974, 100, note 48. The Tanner papers are now held at the University of Liverpool.

125. Van der Waerden 1985; Bashmakova and Smirnova 2000.

126. Cajori 1928; Pycior 1997.

Operations of arithmetic in letters

1. *Operationes logisticae in notis.*

Section (a): On solving equations in numbers

1. *De numerosa potestatum resolutione*, literally 'On the numerical resolution of powers'. Viète's book of the same name begins with finding roots of pure powers (squares, cubes,

etc. up to sixth powers) and then goes on to treat the resolution of 'affected powers', that is, the solution of equations of the form $a^n \pm ba^m = c$ or $ba^m - a^n = c$ where $n > m$ and $b > 0, c > 0$.

2. Viète 1646, 174–175; Witmer 1983, 323–325.
3. Viète 1646, 175–176; Witmer 1983, 326–327.
4. Viète 1646, 176–178; Witmer 1983, 327.
5. Viète 1646, 178–179; Witmer 1983, 327.
6. Viète 1646, 180–182; Witmer 1983, 327–330.
7. Viète 1646, 182–183; Witmer 1983, 331–333.
8. Viète 1646, 183–184; Witmer 1983, 333.
9. Viète 1646, 185; Witmer 1983, 333.
10. Viète 1646, 186–187; Witmer 1983, 333.
11. Viète 1646, 187–189; Witmer 1983, 333–336.

Section (b): On solving equations in numbers

1. Viète 1646, 195–196; Witmer 1983, 339–341.
2. Viète 1646, 196–197; Witmer 1983, 341–342.
3. An *acephalic square* is the case where there are more single digits in the coefficient of the linear term than pairs of digits in the constant term. See Viète 1646, 196; Witmer 1983, 341.
4. Further examples of this problem are provided in rough working on f. 11^V and f. 10.
5. Viète 1646, 197; Witmer 1983, 343.
6. Witmer 1983, 343, translates *artificium parabolae epanorthicum* as 'a corrective device'.
7. Viète 1646, 197; Witmer 1983, 343.
8. Viète 1646, 198–199; Witmer 1983, 344–345.
9. Viète 1646, 199–200; Witmer 1983, 345–347.
10. An *acephalic cube* is the case where there are more pairs of digits in the coefficient of the linear term than triplets of digits in the constant term. See Viète 1646, 199; Witmer 1983, 345.
11. Viète 1646, 200; Witmer 1983, 347–348.
12. Viète 1646, 200–201; Witmer 1983, 348.
13. Viète 1646, 201–202; Witmer 1983, 348.
14. Viète 1646, 202–203; Witmer 1983, 348.
15. Viète 1646, 203–204; Witmer 1983, 348.
16. Viète 1646, 204; Witmer 1983, 348.
17. Viète 1646, 205–206; Witmer 1983, 348–352.
18. Viète 1646, 207–208; Witmer 1983, 352.
19. Viète 1646, 208–210; Witmer 1983, 352.

Section (c): On solving equations in numbers

1. Viète 1646, 211–214; Witmer 1983, 353–357.

2. The canonical form given here is of a new kind, not used in Sections (*a*) and (*b*). Harriot showed how such canonical forms arose in Section (*d*).

3. Viète 1646, 214–216; Witmer 1983, 357.

4. Viète 1646, 216–218; Witmer 1983, 357–362.

5. Viète 1646, 217; Witmer 1983, 358.

6. Viète 1646, 218; Witmer 1983, 360.

7. Viète 1646, 219–220; Witmer 1983, 362–365.

8. Viète 1646, 221–223; Witmer 1983, 365–370.

9. Viète 1646, 222; Witmer 1983, 367–368.

10. Viète 1646, 223; Witmer 1983, 369–370.

Section (d): On the generation of canonical equations

1. Reciprocal equations were also recognized by Viète though the name is Harriot's. See Viète 1646, 219; Witmer 1983, 362.

2. $a = d/3$ satisfies $aaa + daa + bca = bcd$ only if $2d^2 = 9bc$. This is the case in each of the first three examples given by Harriot, but not in his counter-example, $aaa + 6aa + 9a = 54$.

3. The pagination $d.7.2$) indicates that this sheet was a later addition to sheet $d.7$). On this sheet, for the first time in the treatise, Harriot specifically indicated the possibility of a negative root, $a = -f$. Many years later John Wallis quoted the same example, $a = -f$, as an example of Harriot's use of negative roots, and referred to a list in the *Praxis* of equations that could lead to such a possibility (Wallis 1693, 210; *Praxis*, 14). In the *Praxis*, however, negative roots were never mentioned, suggesting that Wallis may have been familiar with the manuscript version of Sheet $d.7.2$).

4. Since it is supposed that $b > 0$ and $c > 0$, the values of d and f found in column 1 and used in the ensuing reduction have imaginary parts. Harriot does not comment on this. Torporley, however, when he wrote the Corrector, specifically mentioned sheet $d.7.2$), and claimed that Harriot's reduction was impossible: '*Si sit possibile ut ex aequatione quadrinomia generetur binomia necesse est ut in gradibus ablatis coefficientes utrumque negativa sint aequalibus coefficientibus affirmativis sed in hisce problematis impossibile ut coefficientes utrinque (hoc est in utroque gradu ablato) negativa sint aequales coefficientibus affirmativas. Ergo, in hisce problematis non est possibile ut ex aequatione quadrinomia generetur binomia.*'; 'If it were possible that from a quadrinomial equation there should be generated a binomial, it is necessary that in the removed terms the negative coefficients of both should be equal to the positive coefficients. But in these problems it is impossible that the negative coefficients of both (that is in both removed terms) should be equal to the positive coefficients. Therefore, in these problems it is not possible that from a quadrinomial equation there should be generated a binomial.' (*Corrector*, f. 11; Halliwell 1841, 114). Torporley did not explain why the magnitudes of the negative and positive coefficients could not be equated, but presumably it was the imaginary quantities involved that for him rendered Harriot's working 'impossible'.

5. In the *Praxis*, Warner omitted this second condition and the working that arose from it on the grounds that it was unclear: '*reductiones tamen earum cum in autographis*

obscurius traditae sint, ad meliorem inquisitionem referenda sunt'; 'the reductions of these, however, since they are treated more obscurely in the manuscripts, must be referred to a better inquiry' (Praxis, 45–46, Problems 19–21). It is true that in the manuscript Harriot's second condition is not separately stated, and is only to be found embedded in the working. Torporley, however, understood the problem better than Warner did: 'Impossibile igitur est, ut ad unicam positam aequalitatem coefficientis partium inferatur ablatio plurium quam unius gradus parodici. Quod ipsum satis erat notum Harrioto. Nam in singulis illis ejus paralogysmis assumit ut in confesso duplicem partium duplicium coefficientium aequalitatem ad binos quosque tollendos gradus.'; 'It is therefore impossible, that by one supposed equation in the parts of the coefficient, there may be carried out the removal of more than one of the lower order terms. Which was itself sufficiently known to Harriot. For in each of those examples, he undoubtedly supposed two equations in the parts of two coefficients for the removal of those two terms.' (Corrector, ff. 11–11^V; Halliwell 1841, 115). Thus Torporley, unlike Warner, was able to identify both the necessary conditions. For Torporley, the stumbling block in Harriot's reduction came later, with the appearance of imaginary quantities, see note 4 above.

6. In this line and in further working, Harriot left his square root signs empty.

7. Sheet d.13.2) is reproduced in Lohne 1966, 195–196. In this sheet Harriot specifically indicated the possibility of an imaginary root, $a = \sqrt{-df}$. Many years later John Wallis quoted $a = \pm\sqrt{-df}$ as an example of Harriot's use of imaginary roots, and referred to a list in the Praxis of equations that could lead to such a possibility (Wallis 1693, 210; Praxis, 14–15). In the Praxis, however, imaginary roots do not appear except to be dismissed as 'impossible'. This suggests that Wallis may have been familiar with the manuscript version of Sheet d.13.2) (see also note 3 above).

8. There are other sheets on this topic scattered amongst the manuscripts; see, for example, Add MS 6783 ff. 203, 399^V, 407^V, 410^V, 414.

Section (e): On solving equations by reduction

1. De resolutione aequationum per reductionem.

2. Apparatus ad genera species et differentias aequationum adventiturum. There are other sheets containing the same or similar material at Add MS 6783 ff. 404, 406^V, 409^V, 412^V, 412–424, 425^V

3. The first table (on p. 175) sets out all possible combinations of signs for equations with up to five terms. Note that in writing down the totals (1, 3, 7, 15, 31) Harriot excluded cases with all negative terms. The second table (on p. 176) summarises equations of each degree. For example, the 4 in the left hand column denotes four types of cubic equation: those with cube term only (1 case); cube and linear term (3 cases, as can be seen from the previous table); cube and square term (also 3 cases); cube, square and linear term (7 cases).

4. This table (p. 177) calculates the totals from the previous table, and therefore gives the total number of cases for equations up to the sixth degree.

5. Torporley placed this and the next four sheets, with the same heading, after Sections (a), (b) and (c), on the numerical solution of equations. However, these sheets,

paginated by Harriot as $e.3.2$) to $e.3.6$), are not concerned with methods of solution, but provide systematic lists of types and cases of equations. They are therefore closely related to Sheets $e.2$) and $e.3$) and, as Harriot's pagination indicates, come naturally at this point in the text.

6. '*species resolutionis*' or 'the letters of the solution'.

7. In the manuscript, Harriot's use of commas to indicate multiplication makes clear the distinction between the cube of a square (bb,bb,bb) and the square of a cube (ccc,ccc)

8. Lemmas 1 and 2 are written out on the back of sheet $e.8$).

9. '*vel immediate via generali*'; it seems that what Harriot meant here was the general numerical method already outlined in Section (b).

10. For the relationship between the roots of a cubic equation and its conjugate see sheet $d.4$).

11. Here Harriot drew a dotted line back to the same equation a few lines previously.

12. The binomial cubic equations are those with only two terms in the unknown quantity; the first three (according to Harriot's list in sheet ($e.2$)) are those with no square term, set out again on this page.

13. Given any equation, it is always possible to write down its conjugate by appropriate changes of sign (see Sheet $d.4$), and indeed Harriot had just done so. However, since a hyperbolic equation has a single positive real root, its conjugate will have a single negative real root. Thus when Harriot claimed that the conjugate equation was not 'possible', he meant that it could not be solved for a positive real root, as he then proceeded to prove.

14. If 'possible' is taken to mean 'has a real positive solution' (see note 13) then Harriot's proof is correct. If, however, e is allowed to be negative, then in the argument for $e < r$ the statement $f < r$ is incorrect, since $f = r - e > r$, and there is therefore no contradiction.

15. Note that in the two examples that precede this statement Harriot has solved the derived cubic, and its conjugate, for all three roots, positive and negative.

16. For Harriot's proof that a hyperbolic equation has no 'possible' conjugate, see Sheet $e.15$) and notes 13 and 14 above.

17. Sheets $e.26$) and $e.27$) are related to Viète's discussion in *De numerosa potestatum resolutione* of cubic equations with multiple positive roots; see Viète 1646, 218–219; Witmer 1983, 360–362. All the examples on Harriot's sheet $e.26$) are the same as Viète's.

18. The main statement is taken directly from Viète, but Harriot has added the condition in brackets. He has also provided a counter-example to Viète's statement, the equation $aaa - 6aa + 11a = 12$. See Viète 1646, 218; Witmer 1983, 360.

19. As in the previous paragraph the main statement is taken from Viète, but Harriot has again added an extra condition in brackets, and a counter-example to Viète's statement, the equation $aaa - 6aa + 12a = 9$. See Viète 1646, 219; Witmer 1983, 361.

20. The statement is from Viète but Harriot has added a correction in brackets; see Viète 1646, 219; Witmer 1983, 361–362.

21. This condition is not found in Viète's discussion but has been added here, with illustrative examples, by Harriot.

Section (f) On solving equations by reduction

1. Here and in the following examples Harriot wrote simply *ee*, but it is clear from what follows that this is to be understood as *aaee*.

2. Harriot's method here is to make the left hand side of the equation *aaaa* = $\mp$ 2*ccda* − *xxxz* a square of the form (*aa* + $\frac{ee}{2}$). The right hand side of the equation then becomes *eeaa* $\mp$ 2*ccda* + ($\frac{eeee}{4}$ − *xxxz*), also a square if (and only if) the condition *ee*($\frac{eeee}{4}$ − *xxxz*) = (*ccd*)^2 is satisfied. The quantity *e* can be found from the cubic equation in *ee* that arises. Once *e* is known, the original quartic can be written and solved as a product of two quadratics, though Harriot here omits this part of the working. The method was originally devised by Ludovico Ferrari and Girolamo Cardano and published in Cardano 1545. There are systematic accounts of the method in Bombelli 1572, 353–411 and in Stevin 1585, 81–87. It is also to be found in Viète's *De aequationum recognitione*, see Viète 1646, 140–148. All these earlier authors were known to Harriot.

3. In this and the following parts of Section (*f*) Harriot wrote some of his equations vertically, and showed parallel arguments side by side. I have rearranged the material in a way that is more easily followed by a modern reader accustomed to reading equations horizontally and to following the development of an argument down the pages.

4. The same equations are treated similarly in Add MS 6783, f. 119.

5. This sheet is reproduced in Lohne 1966, 198–199.

6. Harriot used the word *hypostatic* to describe real roots and *noetic* for what are now known as imaginary roots.

7. In Torporley's *Summary* and in the modern ordering of the manuscripts, sheet *f* [γ].14) (Add MS 6783, f. 157) is preceded by sheet *d*.13.2) (Add MS 6783, f. 156), which introduces the possibility of imaginary roots. Sheet *d*.13.2) could have been placed in this sequence either by Torporley or by Harriot himself.

Appendix

1. *Corrector*, f. 9; Halliwell 1841, 111.

2. Another version of the material in Sheet 4) can be seen in British Library Add MS 6787, f. 567, in an earlier or different hand: the letters are in capitals and an elongated Z is used to indicate equality. In the *Summary* Torporley added a fifth sheet, which he denoted X.5), containing some verse on rules for multiplication of negative quantities: *If more by more must needs make more Then lesse by more makes lesse of more*, etc. This sheet was not, however, included in Harriot's own pagination of this section, and is not directly related to his algebra, and so has not been included; for a facsimile of the original and detailed discussion of the meaning of the verses, see Tanner 1980, 148–150.

3. *Corrector*, f. 9^V; Halliwell 1841, 111.

4. *Corrector*, f. 9^V; Halliwell 1841, 111.

5. *Corrector*, f. 9^V; Halliwell 1841, 111.

6. *Corrector*, f. 9; Halliwell 1841, 111.

7. *Corrector*, f. 9; Halliwell 1841, 111.

8. *Corrector*, f. 9; Halliwell 1841, 111.

Bibliographies

꙳

1. Primary sources: manuscripts

Anonymous, on Aylesbury's books and papers, Bodleian Library MS Rawlinson B.158.

Cavendish, Charles, mathematical papers, British Library Harley MSS 6001–6002, 6083, 6796. Photocopies are kept alongside the Harriot papers in Cambridge, Durham and Oxford.

Harriot, Thomas, mathematical papers (i), British Library Add MSS 6782–6789, with copies in the University Libraries of Cambridge, Durham and Oxford.

Harriot, Thomas, mathematical papers (ii), Petworth MSS HMC 240–241, with copies in the West Sussex Record Office, Chichester.

Harriot, Thomas, *Treatise on equations*, dispersed through British Library Add MS 6782–6783.

Harriot, Thomas, *Examinatio Stifelius de numeris diagonalibus*, British Library Add MS 6782, ff. 84–94.

Harriot, Thomas, *De numeris triangularibus et de inde progressionibus arithmeticis*, British Library Add MS 6782, ff. 107–146.

Hartlib, Samuel, The Hartlib papers, Sheffield University Library, and CD.

Pell, John, mathematical papers, British Library Add MSS 4397–4404, 4407–4431.

Rigaud, Stephen Peter, *Astronomical miscellanea*, Bodleian Library MS Rigaud 51.

Rigaud, Stephen Peter, *Astronomical papers*, Bodleian Library MS Rigaud 56.

Rigaud, Stephen Peter, *Thomas Harriot*, Bodleian Library MS Rigaud 9.

Rigaud, Stephen Peter, *Hadley and Harriot papers*, Bodleian Library MS Rigaud 35.

Rigaud, Stephen Peter, *Rigaud letters, II, F–M*, Bodleian Library MS Rigaud 61.

Torporley, Nathaniel, *Congestor analyticus*, Lambeth Palace Library, Sion College MS Arc L.40.2/L.40, ff. 1–34^{v}.

Torporley, Nathaniel, *Corrector analyticus artis posthumae Thomae Harrioti*, Lambeth Palace Library, Sion College MS Arc L40.2/E10, ff. 7–12.

Torporley, Nathaniel, *Summary*, Lambeth Palace Library, Sion College MS Arc L.40.2/L.40, ff. 35–54^{v}.

Warner, Walter, mathematical papers (i), British Library Add MSS 4394–4396.

Warner, Walter, mathematical papers (ii), Northamptonshire Record Office IL 3422, VI, ff.1–23.

2. *Primary sources: printed books*

al-Khwārizmī, *c.* 825, *Al-jabr wa'l-muqābala* translated in Karpinski 1915; Grant 1974, 106–111.

Apollonius, 1537, *Apollonii Pergei . . . opera*, edited by Joannes Baptista Memus, Venice.

Apollonius, 1566, *Apollonii Pergaei conicorum libri quattor*, edited by Federico Commandino, Bologna.

Aubrey, John, 1898, *Brief lives, chiefly of contemporaries, set down by John Aubrey between the years 1669 and 1696*, 2 vols (A–H; I–W) edited by Andrew Clark, Oxford.

Baillet, Adrien, 1691, *La vie de monsieur Des-Cartes*, Paris.

Baillet, Adrien, 1693, *The life of monsieur Des Cartes*, London.

Bombelli, Rafael, 1572, *L'algebra*, Bologna.

Brouncker, William *et al.*, 1658, *Commercium epistolicum de quaestionibus quibusdam mathematicis nuper habitum*, Oxford.

Brouncker, William, 1668, 'The squaring of the hyperbola by an infinite series of rational numbers', *Philosophical transactions* **3**, 645–649.

Cardano, Girolamo, 1545, *Artis magnae, sive de regulis algebraicis liber*, Nuremberg, generally known as the *Ars magna*.

Dary, Michael, 1664, *The general doctrine of equation*, London.

Descartes, René, 1637, *La géométrie*, appendix to *Discours de la methode*, Leiden.

Descartes, René, 1682, 'The judgement and approbation of Renatus Descartes on this Idea', *Philosophical collections* **5**, 144–145.

Descartes, René, 1897–1910, *Oeuvres de Descartes*, edited by Charles Adam and Paul Tannery, 12 vols, Paris.

Diophantus, 1575, *Diophanti alexandrini rerum arithmeticarum libri sex*, translated by Guilielmus Xylander, Basel.

Ghetaldi, Marino, 1603, *M.G. . . .promotus Archimedis, seu de variis corporum generibus gravitate et magnitudine comparatis*, Rome.

Gibson, Thomas, 1655, *Syntaxis mathematica*, London.

Halliwell, James Orchard, 1841, *A collection of letters illustrative of the progress of science in England from the reign of Queen Elizabeth to that of Charles the Second*, London.

Harriot, Thomas, 1588, *A briefe and true report of the new found land of Virginia*, London.

Harriot, Thomas, 1631, *Artis analyticae praxis ad aequationes algebraicas nova, expedita, et generali methodo, resoluendas: tractatus*, edited by Walter Warner, London.

Harvey, Gabriel, 1593, *Pierces supererogation or a new prayse of the old asse*, London.

Hues, Robert, 1594, *Tractatus de globis et eorum usu*, London.

Kersey, John, 1673–74, *The elements of that mathematical art commonly called algebra*, London.

Leibniz, Gottfried Wilhelm, 1686, 'Treatise of algebra both historical and practical, with some additions, by Iohann Wallis', *Acta eruditorum* **5**, 283–286.

Mennher, Valentin, 1556, *Arithmetique seconde*, Antwerp.
Moore, Jonas, 1650, *Arithmetick in two books*, London.

Nuñez, Pedro, 1564, *Libro de algebra en arithmetica y geometria*, Antwerp.

Oughtred, William, 1631, *Arithmeticae in numeris et speciebus institutio: quae tum logisticae, tum analyticae, atque adeo totius mathematicae quasi clavis est*, London.

Pappus of Alexandria, 1588, *Mathematicae collectiones*, translated by Federico Commandino, Pisauri, reprinted Venice 1589.
Peletier, Jacques, 1554, *L'algebre departie en deus livres*, Lyons, translated as *De occulte parte numerorum quam algebram vocant*, Paris 1560.
Pell, John and Rahn, Johann, 1668, *An introduction to algebra*, London.
Pell, John, 1638, 'An idea of mathematicks', reprinted in *Philosophical collections* **5** (1682), 127–134.
Perez de Moya, Juan, 1573, *Tratado de mathematicas*, 2 vols, Alcala.
Prestet, Jean, 1689, *Nouveaux elemens des mathematiques*, 2 vols, Paris.

Recorde, Robert, 1557, *The whetstone of witte whiche is the seconde parte of arithmetike*, London.
Rigaud, Stephen Jordan, 1841, *Correspondence of scientific men of the seventeenth century*, 2 vols, Oxford, reprinted Hildesheim: Olms 1965 with the same pagination.
Rigaud, Stephen Peter, 1833, *Supplement to Dr Bradley's miscellaneous works with an account of Harriot's astronomical papers*, Oxford, 52 and plate v.
Roche, Etienne de la, 1520, *L'arithmetique ... de la regele de la chose*, Lyon, based on the unpublished *Triparty* of Nicholas Chuquet, 1484, reprinted 1538.

Stevin, Simon, 1585, 'La disme', in *L'arithmetique contenant les computations ... aussi l'algebre*, Leiden.
Stifel, Michael, 1544, *Arithmetica integra*, Nuremberg.

Torporley, Nathaniel, 1602, *Diclides coelometricae*, London.
Torporley, Nathaniel, 1841, *Corrector analyticus artis posthumae Thomae Harrioti*, in Halliwell 1841, 109–116.

Viète, François, 1591, *In artem analyticem isagoge, seorsim excussa ab opere restitutae mathematicae analyseos, seu algebra nova*, Tours.
Viète, François, 1592, *Effectionum geometricarum canonica recensio*, Tours.
Viète, François, 1593a, *Zeteticorum libri quinque*, Tours, reprinted as *Cinque livres de Zetetics*, Paris 1630.
Viète, François, 1593b, *Supplementum geometriae ex opere restitutae mathematicae analyseos seu algebra nova*, Tours.

Viète, Francois, 1600a, *Apollonius Gallus*, Paris.

Viète, François, 1600b, *De numerosa potestatum ad exegesin resolutione*, edited by Marino Ghetaldi, Paris.

Viète, François, 1615a, *De aequationum recognitione et emendatione tractatus duo*, edited by Alexander Anderson, Paris.

Viète, François, 1615b, *Ad angularium sectionum analyticen theoremata*, with proofs supplied by Alexander Anderson, Paris.

Viète, François, 1631, *In artem analyticem isagoge et ad logisticen notae priores*, published by Jean de Beaugrand, Paris.

Viète, François, 1646, *Opera mathematica*, edited by Frans van Schooten, Leiden, reprinted Hildesheim 1970.

Wallis, John, 1685, *A treatise of algebra both historical and practical shewing the original, progress, and advancement thereof, from time to time; and by what steps it hath attained to the heighth at which now it is*, London.

Wallis, John, 1693, *Opera mathematica*, Vol II, Oxford, reprinted Hildesheim: Olms 1972 with the same pagination.

Witmer, Richard T. (translator), 1983, *The analytic art*, Kent State University Press.

Wood, Anthony, 1691–92, *Athenae oxoniensis*, 2 vols, London; page references are to the third edition, by Philip Bliss, 5 vols, London 1813–20.

3. Secondary sources

Bashmakova, Isabella and Smirnova, Galina, 2000, *The beginnings and evolution of algebra*, Mathematical Association of America.

Bos, Henk, 2001, *Redefining geometrical exactness: Descartes' transformation of the early modern concept of construction*, New York: Springer.

Cajori, Florian, 1928, 'A revaluation of Harriot's *Artis Analyticae Praxis*', *Isis* **11**, 316–324.

Cajori, Florian, 1928–29, *A history of mathematical notations*, 2 vols, Chicago: Open Court, reprinted New York: Dover 1993.

Cantor, Moritz, 1894–1908, *Vorlesungen über Geschichte der Mathematik*, 4 vols, Leipzig.

Chemla, Karine, 1994, 'Similarities between Chinese and Arabic mathematical writings: (I) root extraction', *Arabic sciences and philosophy* **4**, 207–266.

Clucas, Stephen, 1991, 'Samuel Hartlib's *Ephemerides*, 1635–59, and the pursuit of scientific and philosophical manuscripts: the religious ethos of an Intelligencer', *The seventeenth century* **6**, 33–55.

Fauvel, John and Gray, Jeremy (editors), 1987, *The history of mathematics, a reader*, London: Macmillan.

Fauvel, John; Flood, Raymond and Wilson, Robin, 2000, *Oxford figures*, Oxford: University Press.

Flegg, H. Graham, Hay, Cynthia M. and Moss, B. (editors), 1985, *Nicolas Chuquet, renaissance mathematician*, Reidel.

Fox, Robert (editor), 2000, *Thomas Harriot: an Elizabethan man of science*, Aldershot: Ashgate.

Gaukroger, Stephen, 1995, *Descartes: an intellectual biography*, Oxford: Clarendon Press.

Grant, Edward (editor), 1974, *A source book in medieval science*, Harvard: University Press.

Hervey, Helen, 1952, 'Hobbes and Descartes in the light of some unpublished letters of the correspondence between Sir Charles Cavendish and Dr John Pell', *Osiris* **10**, 67–90.

Hutton, Charles, 1795–96, 'Algebra', *A mathematical and philosophical dictionary*, 2 vols, London, I, 63–97.

Hutton, Charles, 1812, 'History of algebra', *Tracts on mathematical and philosophical subjects*, 3 vols, London, II, Tract 33, 143–305.

Karpinski, Louis Charles, 1915, *Robert of Chester's Latin translation of the algebra of al-Khowarizmi*, New York: Macmillan.

Klein, J., 1968, *Greek mathematical thought and the origin of algebra*, Cambridge, Mass: MIT, reprinted New York: Dover 1992; includes J. Winfree Smith's translation of Viète's *Isagoge* as an appendix.

Lohne, Johannes A., 1966, 'Dokumente zur Revalidierung von Thomas Harriot als Algebraiker', *Archive for History of Exact Sciences* **3**, 185–205.

Lohne, Johannes A., 1979, 'A survey of Harriot's scientific writings' in 'Essays on Thomas Harriot', *Archive for History of Exact Sciences* **20**, 189–312 (265–312).

Milton, Giles, 2000, *Big chief Elizabeth: how England's adventurers gambled and won the New World*, London: Hodder and Stoughton.

Montucla, Jean Etienne, 1799–1802, *Histoire des mathematiques* (second edition), 4 vols, Paris.

Pepper, Jon V., 1967a, 'A letter from Nathaniel Torporley to Thomas Harriot', *British journal for the history of science* **3**, 285–290.

Pepper, Jon V., 1967b, 'The study of Thomas Harriot's manuscripts II. Harriot's unpublished papers', *History of science* **6**, 17–40.

Prins, Jan Lambert Maria, 1992, *Walter Warner (ca 1557–1643) and his notes on animal organisms*, Doctoral thesis, Utrecht.

Pycior, Helena Mary, 1997, *Symbols, impossible numbers, and geometric entanglements*, Cambridge: University Press.

Quinn, David B. and Shirley, J. W., 1969, 'A contemporary list of Harriot references', *Renaissance quarterly* **22**, 9–26.

Rashed, Roshdi, 1974, 'Résolution des Equation Numériques et Algèbre: Saraf-al-Dīn at-Tūsī, Viète', *Archive for history of exact sciences* **12**, 244–290.

Reich, Karin, 1968, 'Diophant, Bombelli, Viète: ein Vergleich ihrer aufgaben', *Rechenpfennige*, Munich.

Seaton, Ethel, 1956, 'Thomas Harriot's secret script', *Ambix* **5**, 111–114.

Shirley, J. W. (editor), 1974, *Thomas Harriot: renaissance scientist*, Oxford: Clarendon.

Shirley, J. W., 1983, *Thomas Harriot: a biography*, Oxford: Clarendon.

Stedall, Jacqueline A., 2000a, 'Rob'd of glories: the posthumous misfortunes of Thomas Harriot and his algebra', *Archive for history of exact sciences* **54**, 455–497.

Stedall, Jacqueline A., 2000b, 'Ariadne's thread: the life and times of Oughtred's *Clavis*', *Annals of science* **57**, 27–60.

Stedall, Jacqueline A., 2002, *A discourse concerning algebra: English algebra to 1685*, Oxford: University Press.

Stevens, Henry, 1900, *Thomas Hariot and his associates*, London.

Tanner, Rosalind Cecilia H., 1967a, 'The study of Thomas Harriot's manuscripts I. Harriot's Will', *History of science* **6**, 1–16.

Tanner, Rosalind Cecilia H., 1967b, 'Thomas Harriot as mathematician: a legacy of hearsay', *Physis* **9**, 235–247, 257–292.

Tanner, Rosalind Cecilia H., 1969, 'Nathaniel Torporley and the Harriot manuscripts', *Annals of science* **25**, 339–349.

Tanner, Rosalind Cecilia H., 1974, 'Henry Stevens and the associates of Thomas Harriot', in Shirley (editor) 1974, 91–106.

Tanner, Rosalind Cecilia H., 1977, 'Nathaniel Torporley's "Congestor analyticus" and Thomas Harriot's "De triangulis laterum rationalium"', *Annals of science* **34**, 393–428.

Tanner, Rosalind Cecilia H., 1980, 'The alien realm of the minus: deviatory mathematics in Cardano's writings', *Annals of science* **37**, 159-178.

Waerden, Bartel Leenert van der, 1985, *A history of algebra from al-Khwarizmi to Emmy Noether*, Berlin: Springer-Verlag.

Witmer, Richard T. (translator), 1983, *The analytic art*, Kent State University Press.

Index